FORD ENGINE BUILDUPS

Covers 302/351 CID Small-Blocks, 1968–1995

4.6L and 5.4L Modular Engines, 1996–2008

Edited by Evan J. Smith

HPBOOKS

HPBooks

Published by the Penguin Group
Penguin Group (USA) Inc.
375 Hudson Street, New York, New York 10014, USA
Penguin Group (Canada), 90 Eglinton Avenue East, Suite 700, Toronto, Ontario M4P 2Y3, Canada
(a division of Pearson Penguin Canada Inc.)
Penguin Books Ltd., 80 Strand, London WC2R 0RL, England
Penguin Group Ireland, 25 St. Stephen's Green, Dublin 2, Ireland (a division of Penguin Books Ltd.)
Penguin Group (Australia), 250 Camberwell Road, Camberwell, Victoria 3124, Australia
(a division of Pearson Australia Group Pty. Ltd.)
Penguin Books India Pvt. Ltd., 11 Community Centre, Panchsheel Park, New Delhi—110 017, India
Penguin Group (NZ), 67 Apollo Drive, Rosedale, North Shore 0632, New Zealand
(a division of Pearson New Zealand Ltd.)
Penguin Books (South Africa) (Pty.) Ltd., 24 Sturdee Avenue, Rosebank, Johannesburg 2196, South Africa

Penguin Books Ltd., Registered Offices: 80 Strand, London WC2R 0RL, England

While the author has made every effort to provide accurate telephone numbers and Internet addresses at the time of publication, neither the publisher nor the author assumes any responsibility for errors, or for changes that occur after publication. Further, publisher does not have any control over and does not assume any responsibility for author or third-party websites or their content.

FORD ENGINE BUILDUPS

First edition: August 2008

ISBN: 978-1-55788-531-9

PRINTED IN THE UNITED STATES OF AMERICA
8th Printing

NOTICE: The information in this book is true and complete to the best of our knowledge. All recommendations on parts and procedures are made without any guarantees on the part of the author or the publisher. Tampering with, altering, modifying, or removing any emissions-control device is a violation of federal law. Author and publisher disclaim all liability incurred in connection with the use of this information. We recognize that some words, engine names, model names and designations mentioned in this book are the property of the trademark holder and are used for identification purposes only.

CONTENTS

ACKNOWLEDGMENTS

The following editors, writers, photographers and Source Interlink staff members contributed the information in this book: Evan J. Smith, Richard Holdener, David Vizard, Michael Galimi, Steve Baur, Samuel James, Jim Campisano and Luke Magnus.

INTRODUCTION

Since 1964 the Ford Mustang has held a special place in the hearts of car enthusiasts worldwide. The iconic Pony car has been powered by a variety of engines over the last 40 years and the editors of *Muscle Mustangs & Fast Fords* magazine have had their hands in stripping and rebuilding just about every version, from mild to wild.

The current Mustang craze came with the introduction of the 1979 Fox-body Mustang. With it came the 5.0/302 engine, first with a carburetor and then fuel injection in 1985 (on automatic versions). The year 1986 brought the 5.0 H.O. with sequential EFI and a much-improved system. A year later Ford upped the ante to 225 horsepower and a new legend was born.

Mustang fans took this model and modified the 302 in many ways, adding cubes through stroker kits, along with blowers, nitrous and turbocharging. With modern fuel injection and the ability to tune for all drivability conditions, horsepower became limitless. Of course, many owners have added cubes by way of the larger 351-based variant of the 302, which can be stretched out to 450-plus cubic inches these days.

In 1996 the Mustang engine family went modular, with the 4.6L SOHC and 4.6L DOHC engines. Ford has since offered no less than 10 different 4.6 and 5.4 modular versions ranging from 215 hp to 500 prancing ponies.

Ford Engine Buildups is a compilation of the most compelling how-to engine buildups covering the entire realm of late-model engines from 302–347 to 351 Windsors and even the current modular lineup of GT and Cobra engines. Inside, you'll find information of all types of power adders, as well as naturally aspirated combinations for the street and for the strip.

You'll also find interesting and exciting stories on using everything from basic stock parts to the most exotic aftermarket hardware. Making horsepower is what this book is all about and we offer you an inside look at the many projects—one, or perhaps many are sure to fit your needs. *MM&FF* is dedicated to bringing you the best information on late-model Ford performance and hopefully this book will help you prepare your Ford for one of the many car show, drag race or road race events that take place virtually every weekend.

—*Evan J. Smith,* Editor

Chapter 1

BLOCK BASICS

What You Need to Know Before You Build Your First Small-Block

Text and Photos by Evan J. Smith

When it comes to building just about anything, you must start with a strong foundation and go from there. It doesn't matter if you're starting a relationship, a house or a racing engine, the best results will come if you work in this order. Of course, you could take shortcuts, but believe me, that's a surefire path to disaster. I'm sure many of you know what I'm talking about. With that said, I'll place the focus of this article on engine building and more specifically on engine blocks.

The block is not what you'd call a sexy part, nor do many people give it a lot of thought, but it is assigned an amazingly difficult task. It has to support the rotating assembly, route oil to all the critical parts, provide a path for the coolant to flow, and provide a location to mount many critical engine accessories.

While many of us quickly and easily modify our engine to make extra horsepower, you should keep in mind that stock blocks are designed for stock applications. What most people don't realize is that increasing the horsepower and torque output also increases the stress on the block.

While the outside of your engine may be a work of art with shiny valve covers and a flashy intake, inside that engine is a downright hostile environment. If you could look inside a running engine you'd find the crankshaft whipping around wildly, retained only by the main caps and bolts. Attached to the crank's journals are the connecting rods, which sling around on the big ends and drive the pistons up and down on the other end.

Many small-block Ford projects begin with a late-model 302 block like this one. These blocks are plentiful and can be found for around $100. All Windsor blocks use 10 head bolts and have five main bearings with the center as the thrust bearing.

Each cycle begins with the intake stroke. The piston is pulled down the bore and the air/fuel mixture is pulled into the cylinder. Next, the piston is forced up the bore with enough fury to compress the mixture of air and gas, which is sealed in the combustion chamber by the piston rings and valves that are shut tightly against the cylinder head. As the

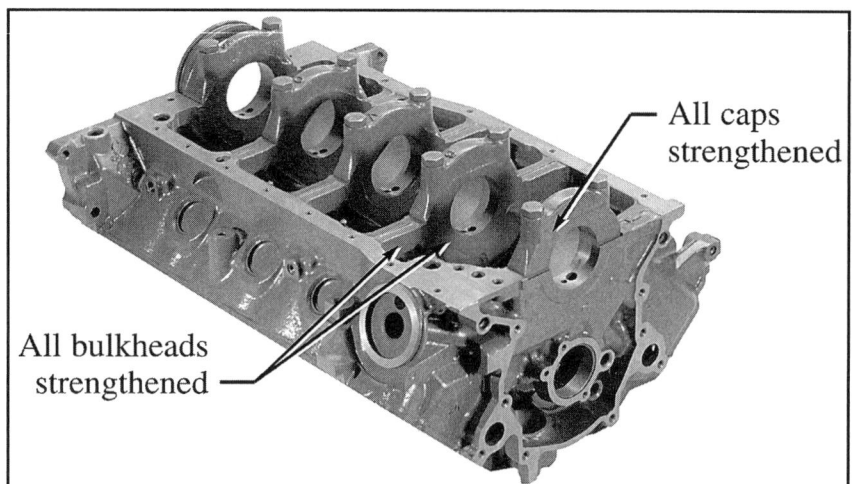

Ford Racing Performance Parts recently released this new Sportsman 302 block. It carries part number M-6010-B50 and is stronger than the standard late-model block. The Sportsman block features two-bolt mains, but has beefier main caps and extra material in the main web area.

If you plan on building an engine for serious competition you should consider upgrading to a racing block. Currently, Dart, Ford Racing Performance Parts, SHM and World Products offer quality aftermarket small-block and Modular engine blocks. Shown are two Ford Racing blocks (left and center) and to the right is a stock late-model 302 block.

Late-model blocks feature two-bolt main caps and all 1985 and newer blocks came with provisions for a roller cam. This is noted by the taller lifter bores and the two bosses to hold the hydraulic lifter retainer (not shown).

This is the lower portion of a cylinder bore in a late-model roller block. When, in 1968, Ford increased the displacement of the 289 to 302, it lengthened the cylinder bores to increase piston stability at BDC.

piston nears top dead center (TDC) the spark plug fires and the mixture ignites then and burns rapidly. In the next moment the piston hits TDC and actually stops, just for a brief moment, though the crank contines to turn. This moment is called "dwell." As the mixture continues to combust, the gasses expand and release incredible pressure and heat energy.

In that same short timeframe the expanding gasses force the piston down the bore and torque is

applied to the crankshaft. At bottom dead center (BDC) the piston halts again before it's rammed up the bore to expel the gasses and to continue the seamless cycle. Meanwhile, the cycle takes place over and over, and it happens faster and faster as rpm is increased. And somehow, this madness is contained by the engine block.

But the engine block has other stresses that we don't often think about. In most cases the engine is bolted to the chassis or to motor mounts, so it's responsible for keeping itself planted between the rails. It also provides a place to mount the engine accessories, such as air conditioning units, pumps and the like, and at the back you'll find the flywheel, clutch/converter and the transmission all bolted directly to the block. Lastly, the block houses other important systems like the oiling and the cooling systems, both of which must work flawlessly to prevent any engine failure.

Aside from simply holding this puzzle together, the block must remain stable under all operating conditions. If the core twists or flexes, the cylinders can become "out of round" and a loss of ring seal

The small two-bolt caps can lead to "cap walk," a case where the main caps move around under load. The scratches indicate that the main cap, which holds the crankshaft in place, has indeed been moving or "walking."

In addition to cap walk we found a crack on the boss for the main journal on this block. At this point the only thing this block would be good for is a boat anchor.

will result. And, if the rings lose their seal you can expect blow-by and a loss in power. In a worst-case scenario, twisting of the block can lead to failures of gaskets and bearings as well as oil and coolant leaks. As you can see, the block serves many functions and picking the right one can mean the difference between owning a reliable and winning engine or one that may come apart on the first run.

A LITTLE HISTORY

The first small-block Ford Windsor engine entered the scene in 1961 when Ford released the 145hp, 221-cubic-inch V-8. In 1963 the engine grew to 260 cubic inches and then to 289 cubes in the mid-'60s. By late 1967 the 289 was replaced by the 302, which remained in production until the mid-'90s. In 1969 a larger version of this engine,

Identification for the late-model blocks can be found on the passenger-side ear at the back of the block. Blocks cast between 1968 and 1981 are often referred to as 302 blocks, while 1982 and newer blocks are called 5.0 blocks. It is said that the early blocks are stronger and better for use in high-performance applications. They can be fitted to accept a late-model roller cam.

the 351 (dubbed the Windsor), was introduced and it benefited from a taller deck height, which allowed for a longer stroke and longer connecting rods. The 351W provided greater torque and therefore proved to be better suited for Ford's larger cars and trucks. The 351W used similar architecture to the 302, but was wider, taller and used a heavier crankshaft with larger main journals (3.00 vs. 2.24 inches). The 351W is not to be confused with the 351 Cleveland, which is yet another Ford small-block with yet another different deck height.

Over the last 20 years Mustang racers have found ways to extract insane levels of performance using many of the stock parts—especially stock blocks, cranks and rods. Unfortunately, pushing these parts too hard leads to engine failures. Still, the parts are plentiful and often not too expensive to replace. The alternative, of course, is to swap your stock block for an aftermarket version and today there are quite a few to choose from.

CHOOSING A BLOCK

For the purpose of this chapter, we'll focus on small-block 302 and 351 blocks, along with a few of the Modular blocks. For late-model Ford owners the most popular engine is the 5.0 or 302-cubic-inch small-block, which came in every V-8 Mustang in 1979 and 1982–'95. This engine is one of the most compact American V-8s ever built and can produce quite a lot of horsepower and torque

On early blocks the cast number is found on the side of the block near the back on the passenger side. Ford blocks have a letter cast into the block that signifies the year of production. Blocks cast in the '60s carry a "C," while blocks cast in the '70s carry a "D." Likewise, "E" was for the '80s and "F" for the '90s. The next number following the letter signifies exact year. So this block is of 1968 vintage.

World Products has developed a line of small-block engine blocks. These new blocks are quite revolutionary, featuring fouradditional head bolts per side and lots of additional innovations.

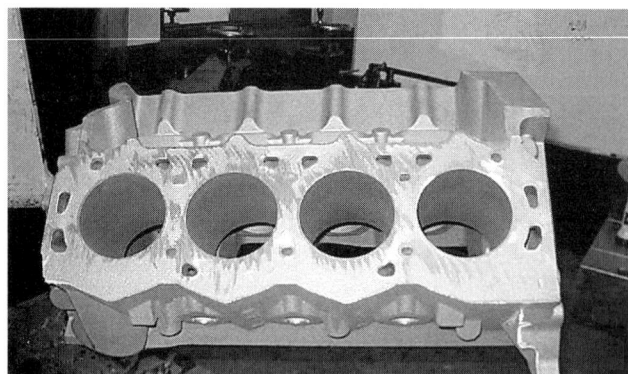

This is a photo of the raw World casting prior to any machining.

World will produce a variety of small-block blocks with all the popular deck heights and crankshaft main journal sizes.

Dart also offers a complete line of aftermarket race blocks in iron and aluminum. Tall deck 351-type blocks are easily distinguished from 289 or 302 blocks by the additional material between the water jacket and the deck.

considering its size and weight.

For most late-model buildups the standard 5.0 engine block will suffice. These blocks are virtually unchanged from the original 302 design, save for a few minor details, and there are many advantages to using one. For starters, they are plentiful and this makes them inexpensive, as late-model 302 blocks can be had for around $50–100. Stock 5.0 blocks do a great job when prepared properly, but there is a limit as to how much abuse they can take before failing. The blocks are cast with a thin-wall designed to save weight and so there is not a lot of rigidity in some of the critical areas. It's virtually impossible to depict at what point failure will occur, but most feel that a small amount of detonation or extended operation above 6,500 rpm will lead to cracks in the main web area and the lifter valley.

"Whether you go with a new or used engine block, the first thing you should do is a visual inspection," stated Charlie Weston of the famed Weston Machine. "Look for the obvious cracks and make sure the freeze plugs are seated properly. If you pressure test the block and there are cracks you'll see bubbles come right out. If the block has been stored inside and has water stains or rust, that is also a good sign of a crack.

This Dart aluminum block can handle a 4.165-inch bore and a 4.250-inch stroke. It has wall thickness of .250 inch and a deck thickness of .675 inch. It also has threaded freeze plugs and can be used with a wet or dry sump oiling system.

"In addition, check all the threaded holes and make sure the oil galleys are clear. Lastly, you want to remove the main caps and look for scratches or marks between the caps and block, check the roundness of the main bearing bore, and make sure the caps snap in and seat properly."

It is possible to improve the strength of a stock block by filling the water jackets with grout or concrete and by adding a main girdle. If done properly the material will support the cylinder walls, thus keeping the cylinders round, which improves ring seal and prevents blow-by. This technique, however, is a debatable one and some engine builders don't feel it's worth doing. Filling the block will not increase the strength of the main webbing. In addition, filling the block should only be done in drag-race applications, because it can

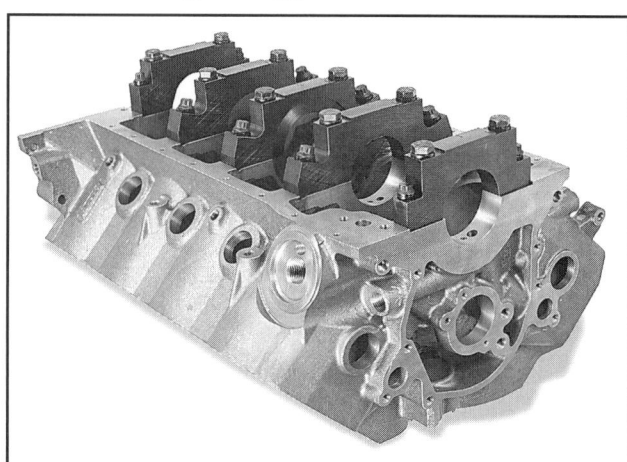

Notice the huge splayed main caps on this Dart 351-style iron block.

Dart's iron block would make a great starting point for just about any competition engine buildup.

Here is a top view of the Dart block. Notice the provisions for the lifter retainer, so the block can be used with a hydraulic camshaft.

Ford Racing's race blocks are also far stronger than stock and have splayed four-bolt main caps.

lead to overheating since no water will circulate around the bottom of the cylinders.

For those looking for a block with increased strength, you're in luck. There is still a good surplus of early blocks and the 1974 and older blocks are said to have a higher nickel content and a thicker wall. There was also a series of blocks cast in Mexico that have thicker main caps and more metal in the web area. These blocks are noted by the word "Mexico" cast into the block. Ford Racing Performance Parts, along with Dart and World Products, offer aftermarket Ford blocks designed for racing or serious street use. Combined there are about 30 different blocks to choose from.

Most engine builders we spoke to agreed that the next level up from a stock block is either the early 302 block or one of the 302 blocks cast in Mexico. In addition, we can't leave out the mighty Boss 302 block, which came with four-bolt mains and screw-in freeze plugs, but good luck finding one of these gems.

The next best option is the Ford Racing 302 Sportsman block that carries part number M-6010-B50. This block features all the stock dimensions as the 5.0-roller 302 block, but has additional material to increase strength in the critical areas, such as the main caps and the cap bulkheads. This block comes machined and ready to use and retails for about $800, making it a great value for the dollar.

For years Ford Racing has offered a complete line of racing blocks for many applications. When it comes to Windsor small-blocks, FRPP has no less than 15 to choose from. The wide range of blocks allows you to mix and match between the variety of deck heights (8.20, 8.70, 9.20 and 9.50 inches), along with the three popular main journal sizes (2.248, 2.749 and 3.00 inches). Between the mix you'll find both iron and aluminum blocks and some designed for wet and dry sump oiling. For a complete list of details regarding Ford Racing blocks please check out the FRPP catalog, check online, or contact a Ford Racing dealer near you.

Another company making racing blocks for Fords is Dart Machinery of Troy, Michigan. Dart has been involved in racing for decades and now offers some of the best castings that racers can get

Compare the additional material in the main webbing area to the earlier photo of the stock block.

The bottom of this stock cylinder bore has been notched for connecting rod bolt clearance. This is necessary when stroker cranks are installed in a stock block.

Most aftermarket blocks come with the notches for stroker kits.

Here is an example of cross-hatching on a freshly honed cylinder.

Though faint, you can still see traces of the crosshatch on this used block. You can also see the ridge created by the piston rings. This is a normal thing to see once an engine is run for some time.

their hands on. Like FRPP, Dart offers a complete line of 302- and 351-based small-blocks with a variety of different dimensions. Dart also sells aluminum blocks and superior iron alloy blocks that can be bored to a maximum of 4.185 inches (on some models). In addition the blocks can be bored for a larger diameter camshaft, some up to 55mm, and all three main journal sizes are offered. The Dart blocks are cast thick to prevent some of the problems with the standard blocks. There is plenty of material in the main web area and cylinder walls that are a minimum of .250 inch.

Perhaps the most innovative block on the market is the brand new Windsor R race block from World Products. This block is state-of-the-art in many ways. "We sat down with Jack Roush and made a wish list of everything we would want in a small-block Ford engine block," stated Bill Mitchell of World Products. "We literally built a block with everything including four extra head bolts for better cylinder head retention." The standard Ford small-block has only 10 head bolts per side (compared to 17 on a small-block Chevy) so there is an inherent problem of the head moving or lifting in high-compression and boosted applications. "There will be four deck heights, three main journal sizes and two bore sizes

available, too" he added.

Some of the other options include billet-steel four-bolt splayed main caps, improved oiling, internal restrictors at front and rear of the block, crossfeeds between lifter pairs to maintain pressure, the dry sump can feed from front or rear of the block and the bellhousing is ribbed for extra strength. These blocks can also accept a 4.200-inch bore, which will make for some big-cube small-blocks.

But World is not the only one casting newly designed Ford blocks. Sean Hyland Motorsports (SHM) of Woodstock, Ontario, Canada, recently released a new line of aluminum Modular blocks designed for endurance and drag racing. "The SHM blocks utilize a 94mm (3.70 inches) bore with a stroke of 105.8 (4.16 inches) or 90mm (3.543 inches) to displace 6.0 (358 cubic inches) or 5.0 liters (305 cubic inches) respectively for each deck height," stated Sean Hyland.

Not only are the blocks larger in displacement, but they have many features than make them superior when compared to stock. For instance, the SHM blocks have a semi-stressed, thicker, capable oil pan rail for dry sump and alternative mounting locations. They also include external oil drain backs, with emphasis on strength and oil return, increased wall thickness of the valley area, shorter water jackets, and a optimized water pump inlet passage for raised water- jacket design. There's also increased head deck thickness, cylinder wall thickness, deep-skirt wall thickness and extra ribbing throughout the block casting.

As you can see, the marketplace for Ford blocks is growing wildly. There are about two dozen aftermarket blocks now available, and one so revolutionary that it will take a new style cylinder head to complete the package. But as Bill Mitchell of World stated, "No one is going to make a head with four extra holes to fit a block that doesn't exist, so we went ahead and built a better block and the heads will come."

This photo shows a simple way to determine the deck height of a block. Be sure to measure from the center of the crank to the deck.

Here is the same tape measure laid across a 302 block.

Weston Machine is one of the leading machine shop facilities in the country. It can perform many tricky machining operations, such as boring a cam tunnel, designing and building aluminum main caps and bushing lifter bores. The latter is done to ensure that all the lifters move on the same plane, thus equalizing cam timing.

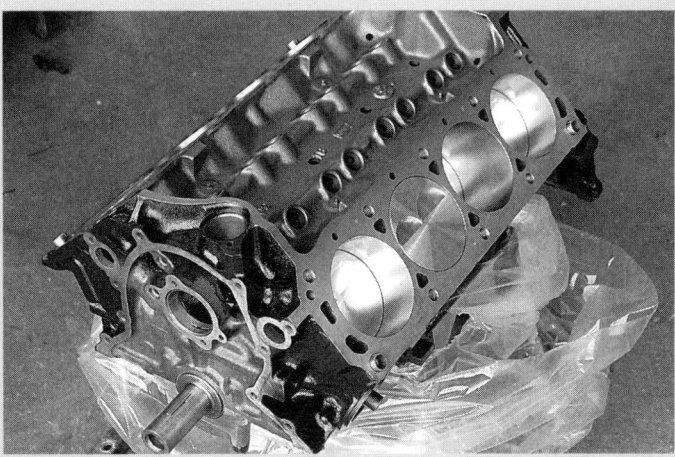

A completed stroker short-block.

This early block has no provisions for a hydraulic roller cam.

A Modular GT short-block built by SHM. Modular blocks use a six-bolt main cap system and a deep-skirted block, meaning the crankshaft sits deeper in the block. This adds rigidity.

This is a rebuilt SHM Cobra short-block.

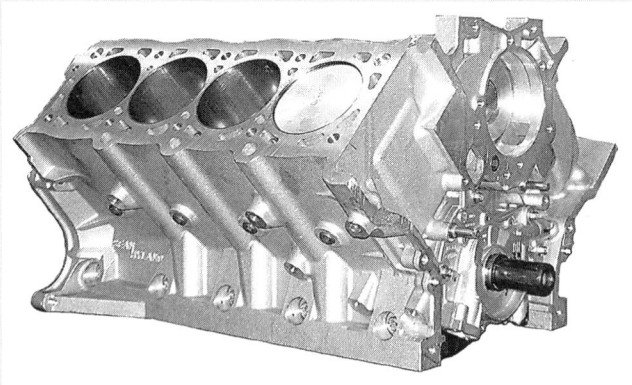

SHM's new aluminum race block is a work of art. Aside from being stronger than stock it will allow for a decent increase in displacement from 281 to 305 on the low-deck models and from 331 to 358 on the tall decks.

All stock Modular blocks have a small 3.552-inch bore.

When preparing a Modular engine for heavy street use or for racing, it's a good idea to install quality aftermarket rods and forged pistons.

A small-block engine getting honed with a torque plate.

Before assembly you should check the squareness of the deck surfaces.

A clean block is a happy block, and this block is about to get a bath in the cleaning tank.

Engine builders use special brushes to clean all the oil galleys and bolt holes prior, and sometimes after, machining.

A main cap girdle can be used to increase the strength of a stock bottom end.

BLOCK GLOSSARY

Bore Gauge: A specially designed dial indicator used to precisely measure the inside diameter of a cylinder bore. Engine builders use a bore gauge to determine the roundness of a cylinder at different locations within a bore. This will determine if the cylinder has any taper and is need of honing.

Bore Spacing: The measured distance between the centerline of two adjacent cylinders. Bore spacing on 302 and 351 engines measures 4.380 inches. Bore spacing on the 4.6 and 5.4 Modular engines measures 3.9370 inches.

Boring: A machining process used to enlarge the bore size of a cylinder.

Cleveland: A specific type of small-block Ford engine produced in the Cleveland, Ohio, engine plant.

Crankcase: The area of the engine that houses the crankshaft and the connecting rods. The crankcase is sealed at the bottom by the oil pan.

Cross Hatch: A machined pattern in the shape of an "X" that is achieved when honing a cylinder. This type of finish allows a small film of oil to be retained on the cylinder walls, yet it allows the rings to seal against the walls during engine operation.

Cylinder Sleeve: A metal sleeve that is used as the cylinder bore in an aluminum block. A sleeve may also be installed in a cast-iron block to repair or modify the original bore.

Deck Height: The measured distance between the centerline of the crankshaft and the deck of the block. Common deck heights for small-block Ford engines in the Windsor family are 8.20, 8.70, 9.20 and 9.50 inches. The 4.6 has a 8.937-inch deck height and the 5.4 has a 10.078-inch deck height. A larger deck height generally means the engine can accommodate a longer stroke and a larger displacement will result.

Decking the Block: The machining process of cutting a small amount of material from the deck surface. This is done to produce a flat and square surface for better sealing of the cylinder head(s).

Deck Surface: The area of the block that the cylinder head(s) mate to.

Dry Sump: A type of oiling system that uses a remote reservoir to contain the oil supply. Dry sump oil systems use an external oil pump to scavenge oil from the crankcase, which reduces windage and generally improves horsepower.

Filled Block: A process of using a concrete-type material to fill up the water jackets surrounding the cylinders. Filling the block in this manner is said to stabilize the cylinders, which can improve ring seal, thus preventing blow-by.

Four-Bolt Mains: Crankshaft main caps using four bolts as a means of retention rather than two.

Head Gasket: A soft material used between the cylinder head and the block to form a pressure seal.

Hone: A machining process where the cylinder walls are cut with a specially designed stone to prepare the cylinder walls before engine assembly. Honing removes a very small amount of material and when done correctly will leave a cross-hatch finish on the cylinder walls.

Long-Block: A common term given to an assembled short-block that has the cylinder heads attached.

Main Caps: Metal caps bolted to the bottom of the block for the purpose of retaining the crankshaft.

Main Girdle: A brace-type devise bolted to the main caps or studs that is used to increase the strength of the main caps, thus limiting any flexing of the main caps.

O-Ringing the Block (Or Heads): A process of machining a small "receiver" groove in the deck of the block or the deck of the heads to accept a special metal ring that helps seal the cylinders.

Short-Block: A term given to an engine block after the rotating assembly (crankshaft, rods and pistons) has been installed.

Siamese Bores: A block with cylinder bores containing no water passages between cylinders. This allows for the casting to have additional material between cylinders to improve strength and to allow for a larger bore size.

Torque-Plate Hone: A method of honing the cylinders using a special plate that is bolted to the head to simulate the cylinder head being installed. This simulates the stress on the block imposed by the head and the bolts and typically produces the roundest bores when compared to honing the block without a torque plate.

Two-Bolt Mains: Type of main cap using only two bolts per cap. All production Ford 289 and 302 blocks came with two-bolt caps except for the 302 Boss blocks.

Wet Sump: A type of oiling system where the oil supply is kept in the oil pan which is attached to the lower portion of the engine.

Windsor: A type of small-block Ford engine (221, 260, 289, 302 and 351) produced in Windsor, Ontario, Canada.

SOURCES

Dart Machinery
353 Oliver St.
Troy, MI 48084
248/362-1188
www.dartheads.com

**Ford Racing
Performance Parts**
15021 S. Commerce Dr., Ste. 200
Dearborn, MI 48120
800/367-3788
www.fordracingparts.com

Sean Hyland Motorsport
691 Jack Ross Ave.
Woodstock, Ont.,
Canada N4V1B7
519/421-2291
www.seanhylandmotorsport.com

Weston Machine
161 11th St.
Piscataway, NJ 08854
732/752-2711

World Products
51 Trade Zone Ct.
Ronkonkoma, NY 11779
631/981-1918
www.worldcastings.com

DIY STROKERS

Building Power by Upstroking Your Engine

**By Evan J. Smith
Photos by the Author and
Courtesy of the Manufacturers**

Gearheads love yapping about hopped-up engines and big horsepower. Speak to us about lopey cams, free-flowing heads or the latest manifold design and we'll listen closely. Chatter about high compression, blowers or a hit of juice and our mouths water. Bench racing is one thing that excites us, and it happens every day in our office. Fact is, in this wonderful world of Ford performance there are dozens of ways to unleash extra ponies; our job is to keep you informed about the hottest setups.

When it comes to building power, one route that's become increasingly popular is stroking, or to be more technically correct, up tro ingWe see it all the time, with 302s, 351s, big-blocks and now with the Modular engines. Racers and street freaks alike know at the end of the day bigger is better—at least when it comes to cubic-inch displacement.

We've learned through trial and error that (all else being equal) larger engines almost always make more power than smaller ones, and by stroking we can find this extra displacement and power.

Unfortunately, there is a right way and a wrong way and there is always the chance that your stroker won't provide the additional horsepower and torque that you've dreamt about. In some cases, a larger engine may even perform worse than the smaller one. Thankfully, in the next few pages we'll reveal some of the myths associated with stroking and we'll dig up a few power secrets to get you ahead of the competition.

BASICS OF BIG

When it comes to engine building there are generally two ways to go big. They are boring and stroking. Stroking is nothing new, but never before has it been easier, cheaper, or more politically correct to do. We call these enlarged engines "strokers" because a stroker engine utilizes increased crankshaft stroke compared to stock, to increase the overall cubic-inch displacement.

Enlarging the diameter of the cylinder bores (read: boring) also increases displacement. It is possible to do both, hence the term "bored and stroked." Of course, you can do just one or the other. If you use this equation: bore x bore x stroke x .7854 x number of cylinders, you can figure out the displacement of any piston-type engine.

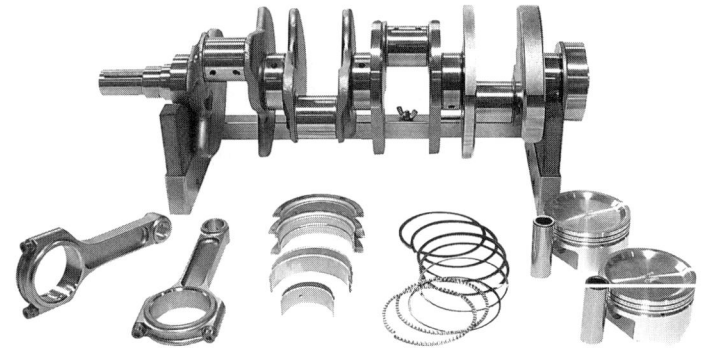

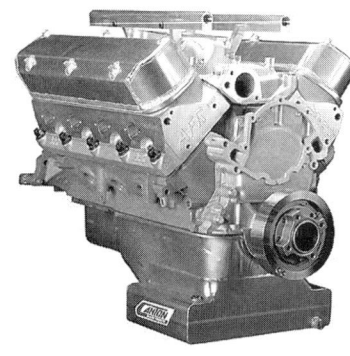

D&D Motorsports Inc. sells a complete line of Hawk-series stroker kits and complete engines.

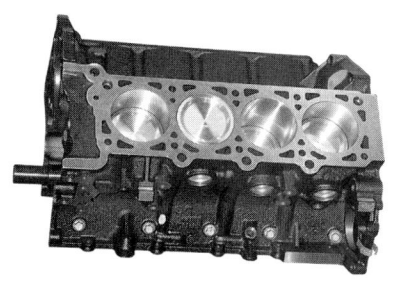

Manufacturers are now building stroker engines and kits for the 4.6- and 5.4-liter Modular engines. Using stock blocks you can take your 4.6 up to 5.1 liters and you can make your 5.4 a 5.8.

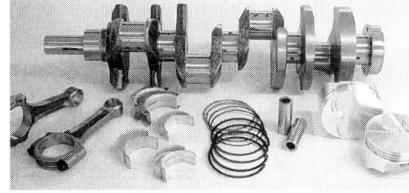

This is a 347 stroker kit offered by Coast High Performance.

You can pay a little or a lot for a stroker crank. The one you decide on should be able to handle the power that you plan to make. This 4340 forged-steel Pro Series crank is from Lunati.

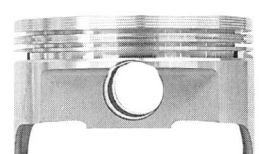

It's easy to see the difference between these CHP pistons. The piston at left is a standard 302 slug and the one on the right is for a stroker. Notice the tighter distance between the pin boss and the top of the piston (this distance is called compression height).

There are benefits to increasing both the bore and the stroke in an engine. Increasing the bore is beneficial for two reasons. It allows for a larger combustion chamber in the cylinder head, which in turn can hold larger valves. And second, it opens up the space between the outer edge of the intake valves and the cylinder bores. This reduces the shrouding of the valves and that almost always improves airflow.

In contrast, lengthening the stroke increases the swept area that the pistons travel in the cylinder bores. This yields a much larger increase in engine size when compared to boring a block. For example, boring a 302 block .030-inch over stock adds a mere 4 cubic inches, making the engine a 306, while stroking the engine by .400-inch (less than half an inch) adds 39 cubic inches. In addition, assuming that the combustion chamber is equal, increasing the stroke automatically increases compression ratio because you'll be increasing the swept volume of the cylinder but not changing the size that the mixture is compressed into.

With any engine there will be a limit to the amount of boring and stroking that can be done. Blocks are cast with only so much wall thickness, which limits bore size. Meanwhile, the stroke is dictated by the deck height of the block and the space available in the crankcase. When stroke is increased, the connecting rod journals are essentially moved further away from the crank centerline. Therefore, you can only increase stroke so far before you create a clearance problem between the connecting rod big ends (namely the rod bolts) and the lower portion of the cylinder bores (on a 302). On the 302 you can gain clearance by grinding the block and this is generally not a big deal.

If we installed a stroker crank with a 3.400-inch arm, the stock connecting rods, which measure 5.09 inches (center-to-center), would cause the piston skirt to crash into the crankshaft counterweights at some point near BDC (the pistons would also slam into the heads at TDC). If you tried to solve the first problem by just putting in longer rods the piston would pop out of the tops of the bores even further and crash into cylinder heads. The solution is then to shorten the overall height of the pistons and go with the longer rods. It is also necessary to relocate the pin bosses closer to the top of the pistons if you want your stroker parts to fit in the block.

Because specialized components must be used, stroker parts are usually sold in kit form. This assures that all the parts will fit and work together. Nevertheless, you can use any combination of oil pan, exhaust, cylinder heads and cam or intake, since the actual engine packaging remains dimensionally stock. And that's one of the greatest benefits of building a stroker. Using a stroker to build displacement means you can keep your current exhaust, stock hood, intake manifold, distributor and engine accessory brackets. This can also make your Mustang or other Ford appear stock as a rock, but with many extra cubes.

With a 4-inch bore and a 3-inch stroke, the 302

CHP offers Tool Steel wrist pins (left) for added strength.

Most stroker kits will include a set of aftermarket connecting rods that are slightly longer than stock. These are from CHP.

Lunati also offers a complete line of connecting rods for stroker applications.

D&D Motorsports offers a line of Hawk stroker engine kits.

Here a block is bored to increase the bore diameter. Both boring and stroking will increase the cubic-inch displacement of an engine.

has the potential to make great power. But when Ford found the need for a more powerful small-block to power its larger cars and trucks, it designed the 351 Windsor, a derivative of the 302, albeit with a raised deck to accommodate a longer stroke. Additionally, the 351 used a beefier crankshaft with larger main journals (3.00 inches compared to 2.248), but the basic block design is the same as the 289/302 and cylinder heads are interchangeable.

In comparison, the 4.6 Mod engine has an almost square bore and stroke. The measurements are 3.552 inches for the bore and 3.5433 inches for the stroke. Modular engines feature smaller bores than the 302 and 351, along with tighter bore spacing. This equates to a shorter overall engine that can fit transversely. Still, the Mod engines have relatively long strokes so they still can make decent torque, despite the smaller displacement.

Another thing to consider is the intended use of the engine, as this plays a huge role in performance, reliability and longevity. Some factors to consider include rpm range, type of use (drag racing, road racing, towing, street performance, etc.) and the expected life of the engine. Strokers tend to make power at a lower rpm, making them great for street use where low- and mid-range torque is king.

MAXIMIZING THE CUBES

In the search for horsepower and torque, engine builders have gone past the standard bore and stroke specifications of the 302, 351 and Mod engines using modified and aftermarket cranks, pistons and rods. With the current list of available aftermarket parts you can now take virtually any Ford engine and make it larger, even the 5.4.

When it comes to building your stroker never lose sight of the big picture. In other words, select parts that support the engine as a whole. Building a 347 is great and you'll love the extra cubes, but those cubes won't be maximized if you're going to use stock heads. On the other hand, installing a set of ported 185cc aluminum heads on a stock block with a stock camshaft will produce the same less-than-stellar results. Either way, remember there has to be a balance between the parts. With a stroker you're adding displacement so you have to complement it with improved airflow through the induction system and the heads.

The foundation for any engine buildup is the engine block. These days you can select between a variety of stock and aftermarket Ford blocks. Always go with the strongest, best block you can afford.

This close-up is of the bottom of a stock 302 block. The 302/5.0 roller blocks feature a thin-wall casting, so they aren't the greatest for making more than 450 hp. Nevertheless, they are cheap and plentiful.

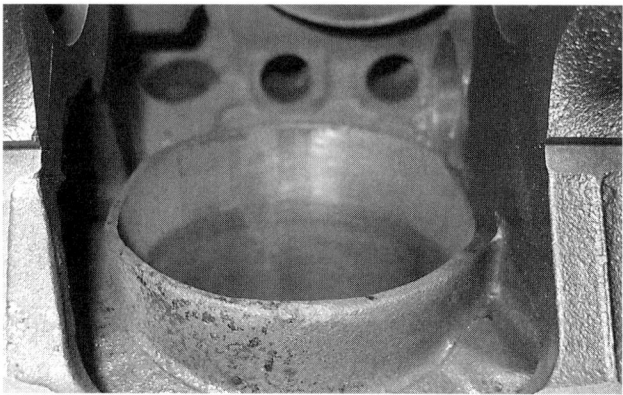

In addition to having four-bolt main caps and more meat in the critical areas, most racing blocks come pre-notched for stroker cranks.

In dealing with 8.2-inch deck height blocks, the most popular stroker kits are the 331 and the 347, however, you can go as large as 360 with the use of an aftermarket block and a 3.50-inch stroke crank. Popular choices for your 351 are 377, 392, 408 and 418. We've seen them as large as 454 with a big bore and a race block. As for the Modular engines, we're seeing 4.6 engines enlarged to 5.1 and 5.2 liters and 5.4s that can measure 5.8 liters or 351 cubes.

It's important to note that increasing stroke has a big effect on piston speed (which is different from engine rpm) and piston speed is something to consider if you plan on turning your stroker at high rpm. Regardless of stroke length, the piston

When installing a stroker it is necessary to notch the bottom of each bore to make clearance for the rods and rod bolts.

must cover the distance between top dead center (TDC) and bottom dead center (BDC) four times per cycle, and when stroke is increased the pistons cover the distance in a quicker amount of time. While this doesn't sound like a big deal, the increase in piston speed places a greater load on the internal engine parts, regardless of whether the horsepower is increased.

For example, imagine that you are a piston, and you will be running (as fast as you can) from bottom dead center to top dead center. Let's say you are inside a 302 and the stroke is 15 feet long. (**Note:** If you've ever done wind sprints you know this will be a daunting task and you will be out of breath after a short time depending on your conditioning.) Now let's say you complete one engine cycle (four total trips) in 6 seconds (we actually did this in the office), and we'll call that 5,000 rpm. Now, let's increase the stroke to 3.400. For this crude example we'll add 4 feet to the stroke for a total of 19 feet. We'll keep the engine (you) turning at 5,000 rpm, meaning that you now have to cover 19 feet in 6 seconds. To do this you will have to accelerate quicker, reach a higher top speed and then decelerate harder. This will undoubtedly place a greater load on your knees and muscles, especially each time you start and stop.

Our subject, an ex-military jumping bean, was in good physical condition yet he was still tired at the end of my experiment. My point is that since the pistons must cover a greater distance in the same time they will be accelerated and decelerated quicker and therefore the load on the internal components is increased. This includes the main webbing in the block, the main caps, wrist pins, the pistons and connecting rods. Knowing this, I recommend using the strongest and lightest pistons and rods when building a high-rpm stroker.

Here you can see just how close the rod bolts get to the block in a stroker.

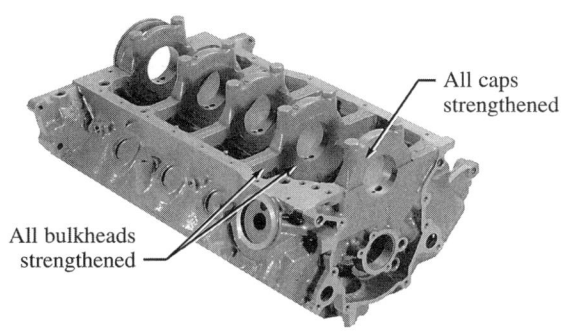

All caps strengthened

All bulkheads strengthened

Ford Racing Performance Parts offer this affordable 302 Sportsman block as PN M-6010-B50. It resembles a stock 5.0 block, but is reinforced in the critical areas.

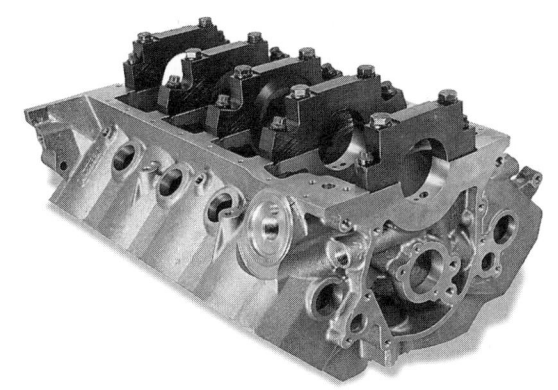

Dart also offers an assortment of aftermarket Ford blocks. And while not shown here, World Products has some nice Ford blocks, too.

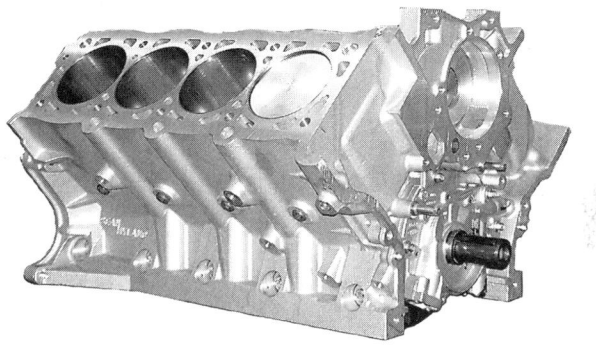

Fret not, Mod fans, SHM offers this aluminum Mod block for serious engine buildups.

When late-model Mustang owners began installing stroker kits more than 15 years ago, there was a big concern with oil consumption. That's because in order to fit the rotating assembly in the block the wrist pin needed to be moved closer to the top of the piston, where it intersected the oil ring. However, over time this problem has proven to be minimal.

"If you want to make torque you can do two things," stated Jim Kuntz of Kuntz & Company in Arkadelphia, Arkansas. "Increase the stroke and shorten the rods. This combination gives you more leverage and greater piston speed to help fill the cylinders at low rpm. You can also work with cam timing to maximize cylinder pressure, as that too will increase power. We use a 3.50-inch stroke with a 5.400-inch rod and a .040-inch overbore on a 302 to make 359 cubic inches. I don't recommend this with a stock block, but we regularly build them with race blocks and they'll turn 8,400 rpm all night long."

Dale Metlika of Propower Performance Parts agreed, "The cars that run best have a terrible rod ratio. We get questions about oil consumption all the time with new customers, but never get any complaints. There is a lot of misinformation out there, but the bottom line is that the larger strokers work because they add displacement. With a stroker the engine will make more mid-range power, even with a stock induction. We've found this to be true especially on the Modular engines. In fact, the Mod engines take less work to stroke."

The only potential problem is with some [Modular] GT blocks, because the stock GT crankshafts don't have counterweights on the center throws and the aftermarket stroker cranks do. Therefore, it may be necessary to have a small amount of machine work done to make the new crank fit.

And any good engine builder will recommend a good oil pan for your stroker, in order to control windage. As crankshaft stroke is increased the throws will be more likely to contact the oil in the pan. This creates drag on the crank and also aerates the oil, causing it to foam. Therefore, we recommend using a deep sump oil pan whenever possible.

BUILDING BLOCKS

While the internal parts make up the stroker, the foundation of your buildup is the block. When it comes to Ford small-block engines there is no

The 4.6 Mod motor is limited to 3.750-inch stroke, because any additional stroke would cause the pistons to come too far out of the bores at BDC.

black magic. There are only so many choices, some stock and some not. It's likely that the block you select will be based on cost, but we recommend using the best block you can afford.

Stock 5.0 Blocks

Stock 5.0 roller blocks are the most common and are great for applications up to (about) 500 hp. Naturally, we've all seen folks make more power with a stock block, way more in some cases, but the fact remains that those blocks are not designed to handle that amount of power. And the reality is that your engine might sustain a healthy life above 500 hp—and it may not. The good news is that there are a few good choices for those looking to make over 500 hp. The next step up is the relatively new Ford Racing Performance Parts M-6110-B50 Sportsman block. This cast is similar to the stock roller block, having two-bolt main caps, but it features increased material in the web and cap areas.

Boss Blocks

Another choice is the Boss block. Boss blocks have four-bolt main caps and are quite strong, but they are rare and hard to find. Thankfully, Ford Racing, along with Dart and World Products offer a complete line of racing blocks for the small-block Ford.

Note: Avoid using a 289 block for any stroker buildup because 289 blocks have shorter cylinder bores (compared to the 302), and this will cause the piston skirts to stick out of the bores at BDC. This can create a stability problem and lead to premature wear of the piston skirts. In addition, the instability of the pistons can lead to a reduction in ring seal resulting in a power loss.

If you're using a stock 5.0 block we recommend that you don't bore the block past .030-inch over. The thin-wall cast is great for weight savings, but lousy for stability and boring too much will weaken the block and can cause a ring seal problem. Whether you use a stock block or a race block we recommend that you have the block Magnufluxed (checked for cracks) and sonic tested if you plan to increase the bore above .030-inch.

CRANK IT UP

If the block is the foundation of your stroker then the crankshaft is the heart. The crankshaft

The bottom-end of this D.S.S. stroker is reinforced with a main stud girdle. Notice that the girdle is notched to clear the rod bolts.

Adding cubic inches means you'll need to improve the induction system, so you might want to consider an aftermarket intake, heads and cam package to complement your stroker short-block.

takes the energy from each power stroke and converts it to torque that's used to drive the car. It also is weighted to balance the rotating assembly and limit unwanted vibration.

Each Ford small-block crankshaft has five main journals on the crank centerline. Those journals keep the crank located properly in the engine block, but also allow it to spin freely on a thin layer of oil between the crank and the bearings. Each crank also has offset journals (four of them in a Ford V-8), which the connecting rods mount to. There are also counterweights opposite the connecting rod journals that offset the weight of the pistons, wrist pins and the connecting rods.

Today, you'll find two basic types of stroker cranks: ones that are modified and those designed and built with the additional stroke. And, you'll find stock or even aftermarket cranks that have the rod journals "offset ground." This requires that the rod journals be ground down to a slightly smaller diameter, and also moved to one side (further out) from the centerline of the original rod journal. This is like putting a smaller circle inside a larger circle, and having the smaller circle set off to one side.

Earlier, we mentioned that the stock strokes are 3.0 inches for a 302 and 3.50 inches for a 351. Common stroker cranks feature strokes of 3.25 inches for a 331; 3.40 inches for a 347; 3.50 for a 360; 3.68 inches in a Windsor 377; 3.90 inches in a 392 and 4.00 inches in a 408. Once you've determined the stroke, you will need to select a crank material. Common crank materials include cast iron, high-nodular cast iron, forged steel and billet steel.

There are a variety of companies selling stroker cranks for Ford engines. The type of crank you select should be matched to the horsepower level. However, it's important to note that supercharged engines place a heavy load on the crank snout so a stronger forged crank may be necessary. Cast-iron cranks are a good choice for street stroker application, but a forged or billet crank will offer the most strength. Again, let price and horsepower level be your guide.

SWINGING THINGS

By now you've realized that any stroker application will require a new set of pistons and connecting rods. Connecting rods are measured from the center of the small end to the center of the big end. Extended rod length is important in a stroker because it moves the pistons further away from the counterweights on the crank. However, lengthening the connecting rods means the pistons must be shorter in height or they will stick out of the bores.

Most of the stroker kits offered for the small-block Ford come with a set of quality "I" or "H" beam rods and pistons with the correct center-to-center dimensions. In addition, there are stock-type rods that will fit dimensionally, but won't be as strong. Since you have to change the piston design, there's no question that the pistons you choose will play a critical role in the performance and longevity of the engine. As I just mentioned, it's necessary to shorten the pistons to make them fit. Shortening pistons in a stroker application is nothing new, but there are things to consider when ordering pistons for your stroker.

We recommend using a forged piston in all stroker applications though the design (dome, dish or flat top) will vary based on your application.

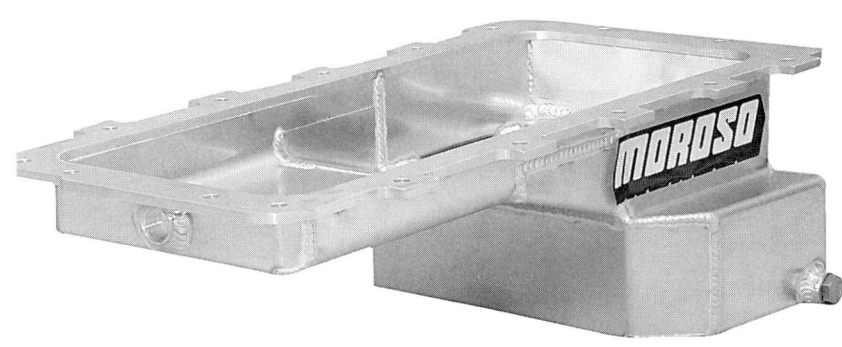

The increased stroke will most likely cause the crank to contact the oil sitting in the pan, this results in drag on the crank and a loss of power. This problem can be helped out with an aftermarket oil pan with a deep sump, like this one from Moroso.

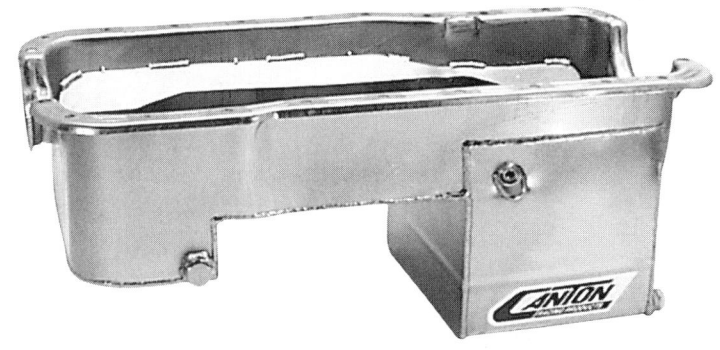

Canton also makes an excellent line of oil pans for Ford engines.

DESTROKING

The last topic for the day is destroking. While it's rarely done in the late-model Mustang universe, destroking is common practice in Competition Eliminator and in some Super Stock Modified classes.

We know that upstroking adds cubic-inch displacement, so naturally destroking reduces an engine's displacement. Destroking is normally found in classes where the rules dictate that a distinct cubic inch-to-weight be maintained. You see, while bigger is generally better, it is generally easier for engine builders to produce a more efficient engine when the displacement is kept small. This is due to the reduction in internal friction and the fact that it's easier to efficiently fill the cylinders in a small engine as opposed to a larger one.

In most cases, a destroked V-8 will feature big cylinder bores and very short stroke. Earlier, I mentioned that a big bore allows for larger valves that are unshrouded, and the short stroke reduces

A windage tray should also be a consideration for any stroker. A windage tray can help to keep the oil off the crank and the horsepower going to the wheels.

piston speed, thus helping to maintain ring seal and reducing internal friction between the rings and the cylinders. Using a short stroke crank also equates to a lighter crank and rotating assembly so the engine can rev quicker and higher. An extreme example of big bore short stroke engines can be found in Formula 1, where a common configuration for the 3.0 liter (181 cid) V-10 might be 3.81-inch bore and a stroke of just 1.59 inches. Amazingly, these engines make upwards of 900 hp and rev to 18,500 rpm. And how about the fact that they make about 4.8 hp per cubic inch—and they're naturally aspirated!

But, remember the application. F1 cars weigh about 1,200 pounds in race trim and they are not expected to accelerate from a standing start (save for the start). In other words, an F1 engine would not do well in a 3,200-pound Mustang—but it sure would sound good.

Instead, we Mustang and Lightning owners need torque, because torque gets the weight moving. And that's where the strokers really shine. The long arm and extra displacement makes torque and torque makes us happy. Just look at the Ford SVT Lightning. The blown 5.4 makes oodles of torque, in most cases more torque than horsepower. And it has stroke, exactly 4.165 inches of it.

SOURCES

Coast High Performance
2555 W. 237th St.
Torrance, CA 90505
310/784-1010
www.coasthigh.com

D&D Motorsports
661/260-2226
www.dndmotorsports.com

D.S.S. Racing
3550 Stern Ave.
St. Charles, IL 60174
630/587-1169
www.dssracing.com

Lunati
4770 Lamar Ave
Memphis, TN 38118-7403
901/365-0950
www.lunaticamshafts.com

Propower Performance Parts
4750 N. Dixie Hwy., Suite 9
Fort Lauderdale, FL 33334
954/491-6988
www.propowerparts.com

SCAT Enterprises Inc.
1400 Kingsdale Ave.
Redondo Beach, CA 90278
310/370-5501
www.scatcrankshafts.com

Sean Hyland Motorsport
691 Jack Ross Ave.
Woodstock, Ontario, Canada N4V1B7
519/421-2291
www.seanhylandmotorsport.com

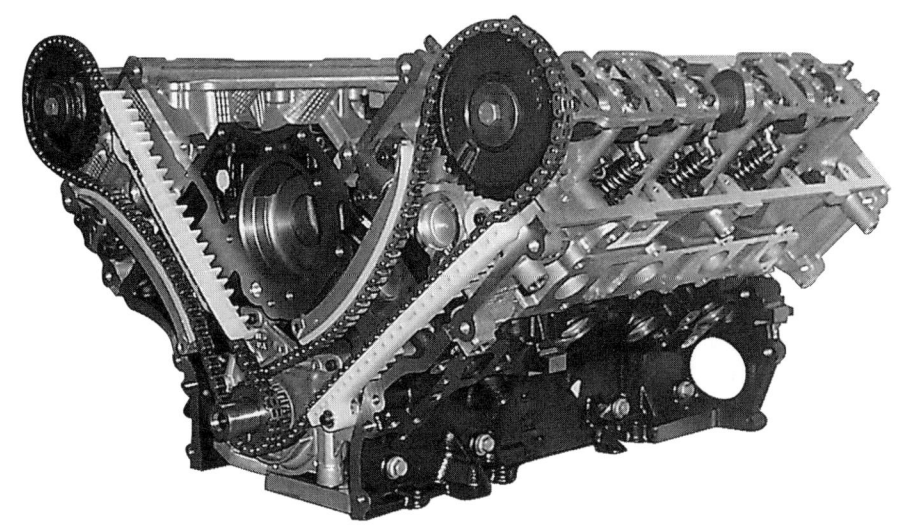

MOD MOTORS

Poked, Stroked and Plenty Powerful

By Steve Baur
Photos Courtesy of the Manufacturers

So you've got the no-torque blues, eh? No problem. *Muscle Mustangs & Fast Fords* is here to help. We've dug up the dirt on stroking the 4.6 and 5.4 Modular engines.

Chapter 2 focused on the science of increasing the stroke of an engine's rotating assembly. That chapter mainly covered the how and why, along with the where, for Windsor engines, so this chapter will focus on who has what for the overhead cam motors.

With more and more engine builders now tweaking the Ford Modular family, there will no doubt be far more than we have assembled here. However, we've contacted some of the oldest, newest, and hottest players in the post-pushrod horsepower era.

When it comes to choosing a new engine for your Mustang, you'll need to consider a great many things other than just power potential. Cost, location, quality of work, and reputation should all be factored into your decision. Above all else, be a smart shopper.

COAST HIGH PERFORMANCE

Coast High Performance offers 5.0 and 5.1 combinations, both of which share the same 3.750-inch stroke. The 5.0 (301 ci) uses a .020 overbore, while the 5.1 (310 ci) takes the cylinders out to 3.625 inches, which is as far as CHP likes to take the factory iron-block cylinder bore. The 5.0 block is preferred for high-boost or high-rpm (9,000-plus) applications, because the extra meat offers a bit more strength than the 5.1.

"The biggest disappointment in the past was cold-start piston rattle," said Chris Huff of CHP. "We've solved that issue by using offset wrist pins to reduce the rod angle. We also offer a thick-wall tool steel pin that is recommended for applications with over 15 psi of boost for high-compression, high-rpm engines."

CHP offers its Street Fighter stroker, which uses a cast crank with a forged connecting rod and piston. Its Pro Street combo steps up to a forged crank, forged 4340 H-beam rods, and forged pistons. CHP sells a main girdle for the 4.6 as well.

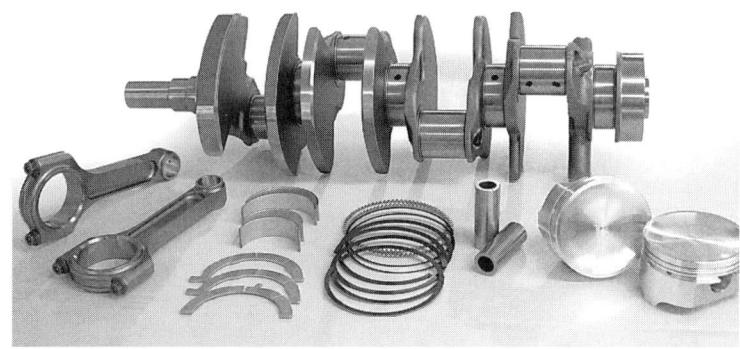

Coast High Performance's Street Fighter 5.1L stroker kit is available with either a nodular iron or a forged crankshaft, depending on your application and budget.

19

Note the raised wrist pin height. This is to accommodate the longer stroke without throwing the piston out of the bore.

MODULAR ENGINE DIMENSIONS

Stock 4.6
Bore: 3.552 inches
Stroke: 3.543 inches
Deck Height: 8.937 inches

Stock 5.4
Bore: 3.552 inches
Stroke: 4.161 inches
Deck Height: 10.078 inches

Used with great success on Ford pushrod motors, the D.S.S. reverse-quench piston maximizes power in a low-compression application.

Despite having an iron block or six-bolt main caps like in the Cobra block, the main caps can still move or walk, especially in high-horsepower applications. Many engine builders offer a main girdle or support bracket like this one from D.S.S. Racing to prevent this from happening, as it can often lead to engine failure.

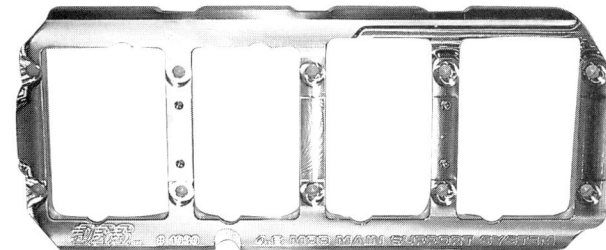

D.S.S. Racing

Coming in at 5.0 is the D.S.S. Modular stroker kit. Using a 3.572 bore and a 3.700 stroke, the kits include forged Pro-Lite pistons and H-beam rods, as well as a forged, purpose-built crankshaft. "The production Cobra cranks are induction-hardened and when they are offset-ground, they lose their strength," said Tom Naegele of D.S.S.

The D.S.S. Modular stroker kit is available as a rotating assembly or as a short-block in Pro Mod or Super Mod forms. Both short-blocks come with the D.S.S. main support, but the Super Mod has a

The D.S.S. stroker package comes standard with a custom forged crank, Pro-Lite pistons, and forged H-beam connecting rods.

40hp advantage over its Pro Mod sibling thanks to the extensive blueprinting of its Level 10 block. "We include our main support because it is extremely effective at controlling the harmonics that cause main cap walk," he said.

On the 5.4 front, D.S.S. only offers a stock stroke overbore Super Mod kit that includes the forged Pro-Lite parts along with the stout forged Navigator crankshaft.

PRO POWER

According to Dale Metlika of Pro Power, "One of the key advantages of a stroker is that you make more power and torque at a lower rpm, which in turn increases engine longevity. We try to sell user-friendly parts, and changing sleeves requires an experienced machine shop, and there's always the possibility of sacrificing strength in the block when you do that."

Leaving sleeves on the shelf then, Pro Power uses a .020 overbore piston (3.572 inches) and combines it with a 3.765-inch stroke to produce its 5.0 stroker package. "Rod ratio is important in certain areas, but it is not the end-all factor. We're mostly concerned with the cylinder head limitations," said Metlika. "Power-adder cars like a lower rod ratio, because they don't need the extra piston dwell to build pressure since the supercharger/turbo is already providing that."

Pro Power has been supplying many high-end customers such as Randy Haywood, whose turbocharged and 5.4-powered outlaw Mustang regularly spins its rotating assembly to 8,500 rpm. Al Papitto races a naturally aspirated 5.4 that also turns 8,000. "You need to look at racing examples that work," noted Metlika.

Pro Power also sells a 5.4 kit. Pro Power actually uses a shorter piston to drop the slug .014 inches

THE CAMMER

Ford Racing Performance Parts offers a few stroker packages for its pushrod motors, but does not offer any for its Modular engines. What they do have is the Cammer.

With the Cammer, Ford opted to use a larger-bore engine as opposed to adding stroke, because it believes the 4.6 is already at its limit for rod-to-stroke ratio for high-rpm usage at 1.674. Ford feels that adding stroke would have made for a risky proposition at high rpm, and the high rpm are needed to take advantage of the great flowing heads.

Ford Racing Performance Parts offers its Cammer crate engine as a larger displacement replacement for the 4.6. At 5.0, it offers 420 hp and 370 lb-ft of torque.

Increasing the bore size was difficult with the small-bore spacing of the Modular engine family (100 mm versus 121.4 mm for the Windsor family of engines). The block has unique thin-wall sleeves and a unique casting that has moved the cooling passages outward from the cylinder bore, which increases the cooling capacity and maximizes cooling efficiency. Additionally, the bulkhead area of the block has been beefed-up to add rigidity and strength, otherwise it shares the same deck height and bore spacing as a Modular 4.6.

The compression ratio is 11:1 (with the Cammer's Ford Racing heads), but will run on 93-94–octane pump gas. Bore is 3.700 inches compared to 3.552 for the 4.6; stroke is the standard 4.6's 3.543 inch. The connecting rods are the standard 4.6 length, and the pistons are unique forged slugs for the Cammer. The crankshaft is also a forged piece, like what's found in the Cobra or manual-transmission Mach 1, but it has been balanced for the Cammer's rotating assembly.

The 5.0 designation has become an icon to Ford owners and enthusiasts in the late- model community. Ford understands the marketing power there and while a 4.9 would have probably got the job done and the 5.1 could have been squeezed out, the 5.0 is the right number. Also, FRPP is casting "5.0" into the side of the block to further differentiate it from the 4.6 block.

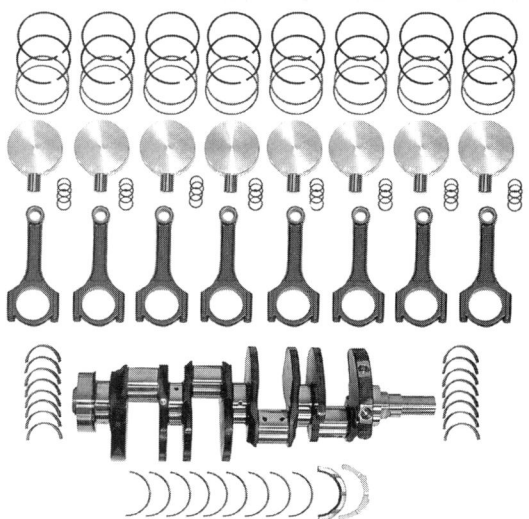

The SHM 4.6 stroker kit comes with Manley forged pistons, I-beam rods, and a forged 3.75-inch stroke crankshaft.

Sean Hyland Motorsports

SHM offers stroker packages in both long- and short-block configurations for two-and four-valve applications. SHM's stroker packages utilize a 3.750-inch stroke crankshaft to raise engine displacement up to 5.0, using an 0.020-inch oversize piston to clean up the cylinder walls during the rebuild process.

"An increase in torque output is the major benefit of using a stroked crankshaft, the trade-off being increased crankshaft loads at high rpm. This is why a matched component rotating assembly is the best way to go when looking into a stroker kit. If you do not have the appropriate connecting rod, bearing, piston, and ring package, engine reliability will ultimately suffer," said Craig English, project design engineer at SHM.

SHM also prefers to use the "Teksid" aluminum blocks found in '96-98 and some '99 Cobras because they seem to be the strongest, but SHM is also nearing production of its own aluminum block casting.

down in the hole for a 4.415 stroke. This, along with the .020 over a 3.572-inch bore, provides for 354 ci. "The combination works out well because it builds a lot of bottom-end torque, which is just what the truck guys need," he said. The 5.4 stroker is optimal for roots/lysholm superchargers because the rpm range is the same. All 5.4 kits come with a billet crankshaft, while the standard 4.6 kit gets a forged crank with a billet piece as an option.

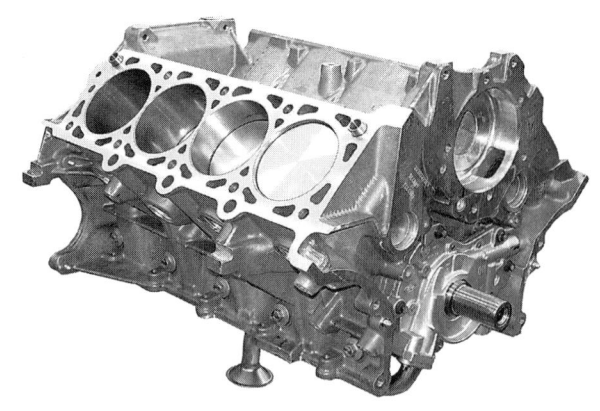

When rebuilding for a high-performance application, Sean Hyland Motorsport prefers to use the '96-98 (and some early '99) Cobra blocks that were manufactured in Teksid, Italy, because they are notably stronger than their later counterparts.

VT ENGINES

VT can take your stock 281 and stretch it out to 302, 304 or even 324 ci. With the 4.6 engine only offering a .030-inch overbore, VT can employ the Darton Modular Integrated Deck to provide for much larger bores.

A dry sleeve is used for naturally aspirated and mild (20 psi or less) supercharged applications, while a wet sleeve MID is employed for high-boost (25-40 psi) applications.

VT uses its own custom pistons made by CP, Oliver billet connecting rods, and 4340 forged crankshafts. VT's 302 uses a .020-inch overbore with a 3.750-inch stroke, while the 304 uses a 3.700 bore with the stock stroke.

"The 304 increases power across the board and offers improved top-end power thanks to the unshrouded valves," said Jim Cushman of VT. "The 324ci motor seems to be the current trend as it offers the best of both worlds with its 3.700 bore and 3.750 stroke. Many of our forced-induction customers opt for the 324 because they want the Darton MID."

As for the 5.4 engine, VT isn't keen on extending the already-long, 4.161-inch stroke. For its short-blocks, VT has the aluminum 5.4 blocks from the Ford GT now available. VT has built a custom 358ci motor with the Darton MID for a turbocharged outlaw car, but that has been the exception.

"We've got a prototype overbored and de-stroked 330ci engine that's designed for better/higher-rpm usage, but it's still being tested before we put it into production. You get what you pay for in this market. There's no reason not to add more cubes when rebuilding, as the cost difference is minimal," said Cushman.

SOURCES
Coast High Performance
2555 W. 237th St.
Torrance, CA 90505
310/784-1010
www.coasthigh.com

D.S.S. Racing
3550 Stern Ave.
St. Charles, IL 60174
630/587-1169
www.dssracing.com

Ford Racing Performance Parts
15021 S. Commerce Dr., Ste. 200
Dearborn, MI 48120
800/367-3788
www.fordracingparts.com

Propower Performance Parts
4750 N. Dixie Hwy., Suite 9
Fort Lauderdale, FL 33334
954/491-6988
www.propowerparts.com

Sean Hyland Motorsport
691 Jack Ross Ave.
Woodstock, Ont., Canada N4V 1B7
519/421-2291
www.seanhylandmotorsport.com

VT Engines
4216 Legacy Pkwy., Suite B
Lansing, MI 48911
519/245-1164

VT Engines uses the Darton Modular Integrated Deck system that replaces the cast cylinders with these steel interlocking sleeves. This enables a much larger increase in displacement than is available when using the stock cylinder arrangement.

Illustration by George Trosley

Chapter 4

331 WAYS TO WASTE AN LS1

The D.S.S. 331ci Stroker Is a Street Fighter

By Steve Baur
Photos by the Author and Courtesy
of the Manufacturers

PART I

Owners of LS1-powered GM cars have become boastful braggarts these days and rightfully so. Even in naturally aspirated form, these engines are terrors from the factory and adding a head and cam package can easily send them to the 500hp mark with the simple turn of a ratchet. You can't do that with any stock small-block Ford, so most of us depend on forced induction to get the job done.

Turning up the boost is essential for making the appropriate amount of ass-whipping power, but in order for your motor to survive for more than just one race, you need to have a good foundation that will hold up to the abuse.

The Package

MM&FF isn't known for doing things half-heartedly, so it should be no surprise when we tell you we're about to take our covert little '90 Mustang GT, drop in a state-of-the-art short-block and turn it into a mind-bending LS1 murderer with help from D.S.S. Racing.

This project car has popped up from time to time and has made mean strides toward eclipsing the quarter-mile in a minimal amount of time. Starting with ProCharger's P1-SC intercooled supercharger, we topped the stock short-block (cam and all) with a wicked set of Brodix M2 ST5.0R cylinder heads and an Edelbrock Performer intake manifold. Breathing through a race-ready set of Bassani stepped long-tube headers and catalytic converters (it did pass New

D.S.S. sells its stroker crank, rods and pistons as a kit for those who wish to build their own motor. With our lofty horsepower goal, we opted to have them stuff these parts in one of their Level 20 CNC-machined engine blocks.

These M2-ported Brodix ST5.0R cylinder heads offered a 90hp gain at the wheels and with their 2.02/1.60 valves, should work well with our increase in displacement, as well as our larger camshaft.

Our ProCharger P1-SC intercooled unit has been impressive from the start, enabling the stock motor to hit 12.20s at over 118 mph on street tires. The same blower put us in the 11.40s at 124 mph after the Brodix heads and Edelbrock intake were installed. It's great to know that such an expensive part of the motor can grow with the car's performance.

The cylinder is bored .025 over, leaving .005 to hone with torque plates.

WHY THE 331?

Why did we pick a 331-inch motor over many of the other popular stroker kits you may ask? We posed the same question to Tom Naegele of D.S.S. and here's what he had to say:

"Our 331 offers the best combination of rod ratio, piston design and ring package for the given 8.2-inch deck height engine. Some people don't consider the frictional losses, poor ring seal and compromised piston design that the larger strokers create. You can fix the problems associated with bigger strokers by using a taller 8.7-inch deck FRRP block, but that incurs more cost. A 331 is almost 40 hp better at 6,000 rpm than a 347. With the 331, we use a 5.315-inch rod, which allows for better ring placement on the piston and proper space between the top and second ring.

"The 347 is good for heavier cars where extra low-end torque is needed. Early on, many builders were using (and some still use) the wrong piston ring packages. This gave this combination a bad reputation for consuming oil. In some extreme cases, the poor rod ratio can even collapse cylinder walls as it side loads the piston in the cylinder really hard. But a 347 is good for 15–20 ft-lb more torque at 3,000 rpm.

"The 327ci stroker is a 331 with a standard bore. It's not a good choice for big-valve cylinder heads due to valve shrouding. The .030-inch overbore of the 331 unshrouds the valves and the machined block offers torque plate honing that a stock block wouldn't normally have. Both of these attributes are worth up to 40 hp. It's really a non-engine builder's combo as anyone with a toolbox can put one together. But you leave a lot of power on the table.

"The 318/320 was an economical combination we used to build when modern strokers were still very expensive, but they're kind of obsolete now. When you get into the 355, bigger is not always better. Rod ratio, piston design and ring seal are traded for extra cubic inches. Some builders don't think about what they are giving up to obtain those cubic inches—usually around 40 to 50 hp and some reliability. This kit is not very popular anymore."

With that said, the 331-cubic-inch stroker package made sense to us since we would be using a stock-block. D.S.S. has made great strides in getting the factory iron to survive under three times the stock power output, but it has its limit like anything else.

Jersey's rigid emissions testing), the LaRocca's Performance–tuned pony has pumped out 473 hp and 476 lb-ft of torque to the wheels, enabling the Midnight Blue Metallic Mustang to run 11.40s at 124mph.

That's some serious street power for sure, but just barely enough to keep up with some of the faster LS1 cars. We could simply add more boost, but we have made this considerable amount of power using the stock 140,000-mile short-block. Let's just say, we felt we were on borrowed time. We called up D.S.S. Racing to find out what they could do for our project car.

The Power Plan

We don't want to just beat LS1s, we want to erase them from memory, and watch them evaporate from our rearview mirror. Thus, we've decided to add a few more cubic inches and get the air/fuel mixture moving faster with a high-performance camshaft. In part one of this buildup, we'll cover the core parts and the machine work that is required to assemble a 331ci stroker engine.

We've decided to stroke the 302 for the increase in displacement, which allows a greater volume of air and fuel to be pumped into the combustion chamber. A longer stroke also has a mechanical advantage by increasing the arm or lever of the crankshaft, which produces more torque. Our new bumpstick will offer far more valve lift than the stock .444-inch we are using now, not to mention more duration.

Part two of our 331 buildup will cover the

The CNC program machines an offset cylinder chamfer for a 4.100 bore. This 60-degree offset chamfer unshrouds intake and exhaust valves, similar to the Boss 302 and 351 Cleveland. Most aftermarket heads are 4.100 bore, as are the head gaskets. A 4.030 bore block without this chamfer leaves a wall for air and fuel to get around on the side of the valve. This subtle machine operation is one of the horsepower enhancements available with a fully CNC-blueprinted block.

The Reneshaw Probing system finds the center of the freeze plug hole so it can be accurately thread milled for the thread-in freeze plugs. The probe also enables accurate location of the head dowel holes. This is what Ford based all their machining coordinates from.

With production blocks that see such high-horsepower levels, using thread-in freeze plugs not only strengthens the block, but also prevents leaks brought on by movement of the block itself. Thread tapping is extremely difficult by hand and the CNC thread-milling program allows more efficient tapping in a shorter amount of time.

installation and testing of the new powerplant in our innocent little Fox. More boost, more cubes and more cam. It should be fun.

The Foundation

D.S.S. Racing was one of the first companies to market Ford crate motors and is probably best known for its Bullet short-blocks, but its latest acquisition, a Haas Horizontal Machining center (more commonly known as a CNC machine) has enabled the company to take engine block modifications to the next level.

This machine does it all, from boring to decking to thread milling and stroker clearance. It also does it perfectly every time. "D.S.S. engines have a reputation for making more power than people expect," says Tom Naegele, Vice President of D.S.S. Raacing. "This is a result of extensive block preparations and ring seal. Ring seal is the most important part to making power."

D.S.S. still offers its Bullet short-blocks, but for higher horsepower applications, the CNC mill works overtime to pump out the Level 10 and Level 20 CNC production blocks.

All of the engine cores that D.S.S. uses are provided by guaranteed core suppliers, but D.S.S. thermal cleans and shot-peens them to stress relieve them. Then they are Magnafluxed to double check viability.

The 10 series is rated at 600 hp while the 20 series piece is good for 675 hp. In addition to the numerous CNC-milling procedures that are completed on the Level 10 piece, the Level 20 receives thread-in freeze plugs that strengthen the sides of the block.

For the enthusiast who needs even more, D.S.S. offers its Street Renegade Bullet short-block, which uses a Ford Racing R302 or optional Dart block, which is good to over 1,000 hp.

Components

For the 331ci motor, each cylinder receives a

The top press-in oil galley plug is a common source of trouble in high oil pressure applications (high-volume pump, 20W50, cold weather, etc.). The CNC machine gives D.S.S. the ability to machine any place on the block making threading this hole for a plug a snap.

The CNC machine makes engraving easy. Other simple mods like this that are gained by using the CNC mill are passed down to the consumer at little or no cost.

The Hines Hard Bearing balancer ensures racing tolerances to plus or minus 1 gram. The 331 stroker kit is externally balanced to 28 ounces.

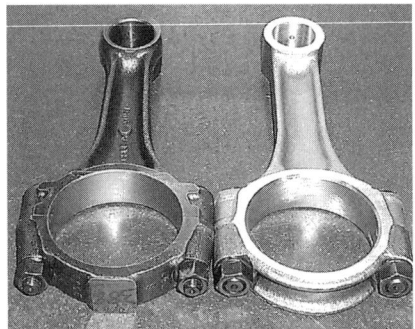

Here you can compare the race-prepped, forged I-beam rod (right) to the stock connecting rod on the left. The D.S.S. pieces come bushed and with ARP hardware.

The D.S.S. max quench reverse dome low-compression piston is an efficient design that weighs about 160 grams less than a stock piston. The lighter the rotating assembly is, the quicker the motor can rev.

Here you can see the installed thread-in freeze plugs. These will add strength to the side of the block.

The Pro Lite pistons use forced pin oiling rather than splash oiling. Another key feature is the sealed power ring package. The low-drag ductile iron rings use thinner top and middle pieces, yet prevent oil consumption.

The trick to clearancing blocks is to take just the right amount of material without taking too much or too little. The CNC machine does just the right amount every time. This gives you the strongest possible block with no chance of interference.

Note the raised wrist pin height in the Pro Lite piston. This is to compensate for the increased stroke length of the crankshaft.

This is the finished short-block with main support installed.

4.030 overbore to be combined later with the 3.25-inch stroke of the nodular iron crankshaft. "We provide a well-engineered high-performance assembly which offers great value for the money," says Naegele. "We like to fit people into combinations that make sense to them. We offer an optional 4340 forged crank that's good up to 1,000 hp, but it's 15 percent heavier and more expensive, which is why the nodular iron makes better sense for the individual who is using a stock-block."

The connecting rods used in the 331 assembly are 5.315-inch forged I-beam pieces that feature bushed ends, full-floating wrist pins and ARP 3/8-inch Wavelock bolts. The piston of choice is D.S.S.'s own Pro Lite forged aluminum unit. Made in-house, these pistons utilize a max quench reverse dome (low compression) for better efficiency and flame travel. Rather than having just a big dish in the top of the piston, the D.S.S. reverse dome is like a mirror image of the combustion chamber itself.

Once the CNC machine is done with the block, assembly is no different than any other 302, but all D.S.S. engines are built by professionals and not your cousin's friend's uncle. "In order to ship motors all over the world like we do, they have to be right the first time," says Naegele. "We have a redundant triple quality control system in place that checks clearances and torque specifications using precision dial bore gauges and micrometers."

The D.S.S. 331ci SuperPro stroker with the Level 20 upgraded block sells for $3,539.95 and is capable of making over 700 hp, although D.S.S. recommends limiting it to about 650 with the stock-block. At this price, the 331 is a good value for the money, and is exactly what we need to support our induction system.

Now that we have a solid bottom end, we plan to crank up the boost on our LS1 killer and go

We turned to Crazy Horse Racing and LaRocca Performance to help with our installation.

Glenn disassembles our trusty 5.0-liter, as we will be reusing many of the existing parts including the cylinder heads, intake manifold and timing chain. This workhorse served us well for 143,000 miles, and delivered 473rwhp.

hunting. Now let's bolt this baby in and hit the dyno and dragstrip.

PART II

To help get our 331 D.S.S. stroker ready for the road, we turned to Crazy Horse Racing in South Amboy, New Jersey, and LaRocca's Performance in Old Bridge Township, New Jersey, for help.

Tom Naegele and Jimmy LaRocca both agreed that 520 rwhp (600 flywheel) is about the limit if you want the engine to last. Our 331 is certainly capable of making much more power simply by turning up the boost or changing the cam and intake manifold to a more supercharged-specific application. However, we intend to keep our pony completely streetable and have no intention (at this time) of installing a rollbar, so we don't see the need to push the envelope. And 500 lb-ft of torque is a lot of power to put down to the asphalt and a task not easily accomplished. With a clear vision of

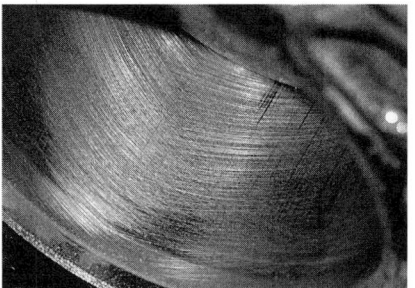

Despite the mileage, all of the cylinders still bore the original cross-hatch marks.

D.S.S. includes a new cam plate, as well as dowel pins for the cylinder heads.

Note how the D.S.S. main support girdle is clearanced for the connecting rods. All bolts are triple checked to ensure your motor is ready to go when you receive it.

The oil pressure sending unit must be reused, and we'll also be using the stock oil pump pickup. You'll also need a new pilot bearing/bushing.

Here you can see the D.S.S. reverse dome low-compression piston. We opted for the O-ringed block to make sure we keep the cylinder pressure where it needs to be.

For our camshaft, we chose a Comp Cams XE274HR grind. This exact cam (PN 35-518-8, $234.95) was used with our Brodix heads in our "Ultimate Guide to Cylinder Heads" (*MM&FF* November 2003) and produced 434 hp in a naturally aspirated 331-inch engine.

We had only logged 140-some miles on the Comp Cams double roller timing chain that we installed when we swapped the Brodix heads on, so we reused it.

We also reused our ARP head studs. For head gaskets, we chose the factory graphite pieces. According to Tom Naegele at D.S.S., they've seen great power had with graphite or Cometic gaskets. Obviously the SuperPro Bullet's straight deck and O-rings help keep everything in its place.

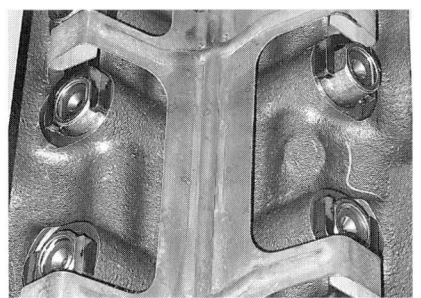

Considering the mileage, the stock roller lifters had served their duty well, and despite not showing any real signs of wear, the factory pieces were traded in for a fresh set from Comp Cams (PN 851-16). The lifter cage was cleaned and bolted to the new engine.

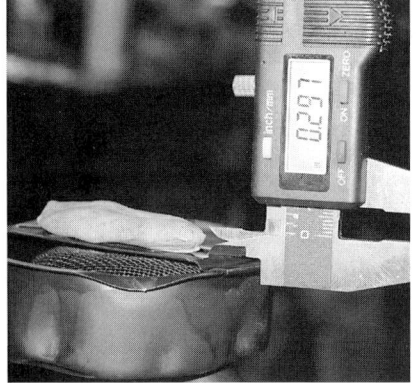

Pickup-to-pan clearance needs to be checked to ensure that there is 3/8–7/16-inch gap. Modeling clay and a micrometer are useful in this operation.

The Brodix M2 ST5.0R heads were installed and torqued to 80 ft-lb. We let them sit for 30 minutes and then backed them off and retorqued them to 85 ft-lb. to make sure there was no more stretch.

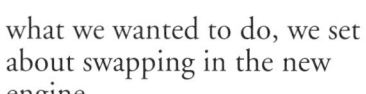
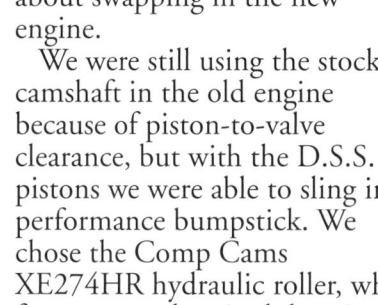

We ordered a new stock oil pump as it offers plenty of petroleum for our pony's powerplant. The stock pickup was cleaned and reused.

what we wanted to do, we set about swapping in the new engine.

We were still using the stock camshaft in the old engine because of piston-to-valve clearance, but with the D.S.S. pistons we were able to sling in a performance bumpstick. We chose the Comp Cams XE274HR hydraulic roller, which features an advertised duration of 274 degrees intake, 282 degrees exhaust, and duration at .050 is 224 degrees intake, 232 degrees exhaust. Gross valve lift is .555-inch intake/.565-inch exhaust and the lobe separation is 112 degrees.

It's designed for an rpm range of 2,200–6,200 rpm and it is the same cam that Richard Holdener

We employed this high-performance chrome-moly ARP oil pump shaft (PN ARP1547904) instead of the stocker. Some prefer to use a stock shaft because it twists when the oil pump seizes, saving the cam gear. We'd rather beef everything up just to be safe.

We called up Holcomb Motorsports in Lumberton, North Carolina, for a new harmonic balancer. This Engine Works unit by Romac (PN NIC208SA) is SFI-approved, fully degreed and good to 8,000 rpm. It features a 28-ounce balance to work with our D.S.S. crankshaft.

used in an engine package that was a 10.25:1 compression 331 with our Brodix ST5.0R heads. With a carburetor, the combination made 434 hp and 437 lb-ft of torque.

It's not necessarily the best blower camshaft, but we went with Holdener's theory that the best supercharged engines are maximized in naturally aspirated form. This allows you to use less boost to attain the same or better performance. Plus Richard has run the same camshaft in his street car for quite a bit of time and was happy with its drivability and performance. That meant a lot to us since ours is primarily a street car.

When you get to this power level, you'll find that there are a lot of ancillary items that need to be replaced with better pieces. Even at stock power levels, the factory engine mounts are easily taxed and if your ride is a high-miler like ours, chances are they're completely shot. We called up Holcomb Motorsports in Lumberton, North Carolina, to get a set of Energy Suspension polyurethane mounts. We wanted them to be strong, but we didn't want to resort to a solid mount.

Holcomb also provided us with a new harmonic balancer/dampener. The Engine Works piece is

Oil leaks suck plain and simple, so we splurged for this one-piece oil pan seal from the auto parts store.

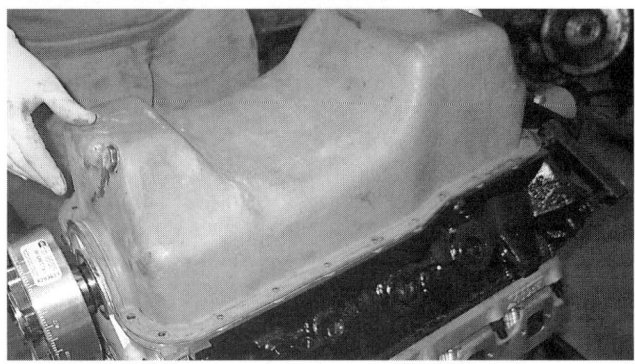

D.S.S. recommends using this '90-earlier oil pan with the flat middle. Some later pans feature an angle in the middle that may contact the main support system and is bad for windage because of its proximity to the crankshaft.

We reused our Comp Cams Magnum 1.6:1 roller rockers.

The intake manifold was bolted up and then it was time to change the injectors.

Since we didn't really want to install an aftermarket fuel system like the one from Aeromotive, LaRocca's Performance recommended stepping up from 42 lb-hr to 50 lb-hr injectors, so we called up MSD for a set of eight squirters. (PN MSD-2013).

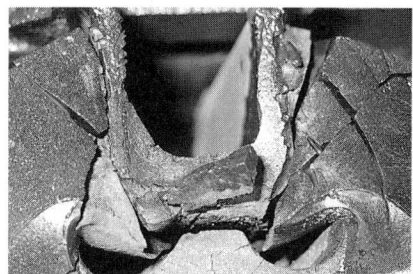

Engine mounts are a common problem with 5.0-liters and regularly fail from the torque of stock engines, let alone supercharged ones.

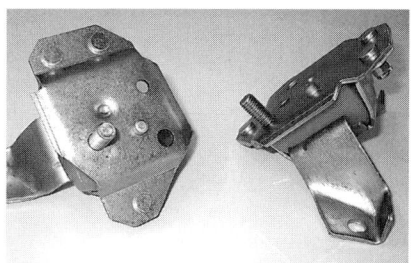

Holcomb Motorsports sent us these Energy Suspension urethane motor mounts (PN ES41122R, $139.95) to make sure our powerplant stays put.

made by Romac and is SFI approved. Our stock balancer was still holding together, but with the considerable investment in the short-block, we didn't want to skimp here.

Another item that we decided to upgrade was the bellhousing. Holcomb shipped us a Lakewood blowproof steel housing that is also SFI approved. It's a little overkill for our "street car," but you can't powershift without your ankles.

Of course, gaskets, a new oil pump, paint for detailing the engine, oil and filter and coolant are all necessary to complete the build. Crazy Horse Racing in South Amboy, New Jersey, handled the engine assembly and installation and had the Mustang up and running in no time.

While Crazy Horse is fully capable of tuning our pony, LaRocca's Performance has been tuning our ProCharged pony since day one, and therefore we already had baseline horsepower and torque numbers from the Dynojet there. That being the case, we headed back to have LaRocca check the tuneup prior to making some dyno pulls.

LaRocca recommended that we step up in injector size from 42 lb-hr to 50 lb-hr. MSD provided us with a set of its competition 50 lb-hr injectors that we complemented with a 3.5-inch Pro-M Univer mass-air meter. The Univer was

designed with the ProCharger's blow-through nature in mind and can be calibrated to your particular injector size when ordered. The computer, mass-air meter and injectors all work in concert to provide the engine with the appropriate amounts of air and fuel, so it is essential that they all know each other.

With the new injectors and meter, we found that the computer had a hard time opening the injectors properly to provide the desired idle. Whereas the '94-up tuning software allows us to adjust the injector's minimum pulse width, the '93-earlier software does not offer that option. To remedy this, we reinstalled the 42 lb-hr injectors and compensated with added fuel pressure (50 psi instead of 43 psi).

LaRocca was indeed right about us needing bigger injectors as the 42s begin to max out around 6,000 rpm. The steady 11:1 air/fuel ratio begins to head north towards 12:1 by that time, but it is still in the safe zone.

With the air/fuel ratio to his liking, LaRocca began tuning for power. At 10 psi of boost, our previous peak dyno figures came in at 473 hp at 5,500 rpm and 474 lb-ft of torque at 4,700rpm. This was with 4 degrees of timing retard and the same 11:1 air/fuel ratio. With zero retard and the same boost, the Mustang laid down 512 hp at 5,800 rpm and

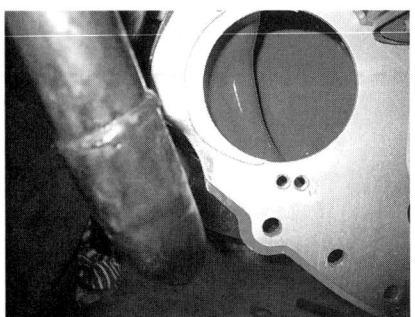

Long-tube headers generally have a clearance problem with the bellhousing, which is one reason why most race headers are fabricated. Cutting the bell housing technically negates its SFI rating, but we're more worried about our ankles, not the regulations. Glenn used a plasma cutter to notch out the metal on the passenger side and then ground down the rough edges for a finished look.

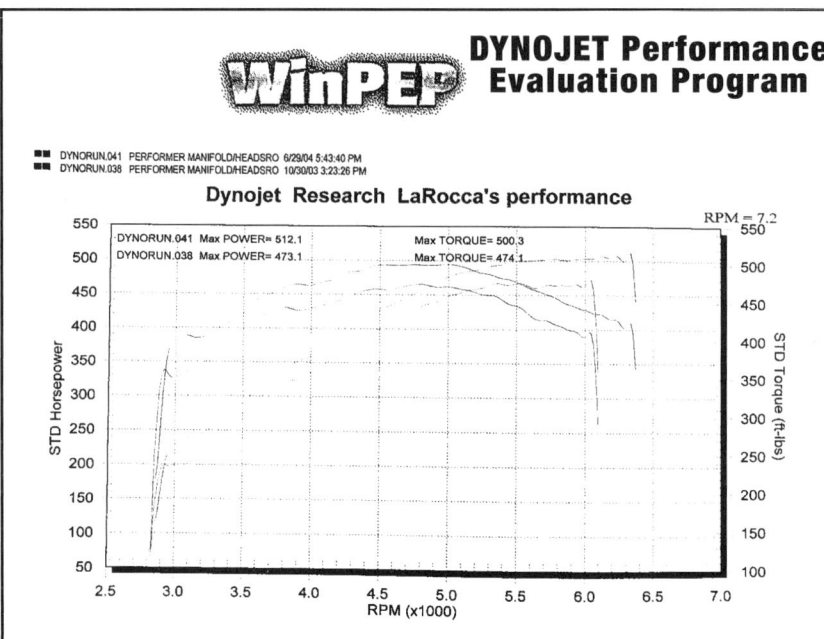

With our boost pressure (10 psi) and air/fuel ratio (11:1) staying the same as before, we picked up 40 hp and 24 lb-ft of torque at the rear wheels. The conservative tune (22 degrees total timing) will keep our engine making loads of LS1-killing power for years to come.

When you change the fuel injector from one size to another, the mass-air meter must be recalibrated as well. We took the opportunity to upgrade from our Pro-M 75mm Bullet to this Pro-M Univer meter. The advanced element placement is designed to read more accurately in blow-through applications like our ProCharger setup.

500 lb-ft of torque at 4,700 rpm. In a 3,100-pound car with a good suspension, that's 9-second power, folks.

As we mentioned before, this combination is easily capable making another 80–100 hp with a little more boost and maybe a different camshaft. But, we're fairly certain we've got enough to take care of those LS1 Corvettes, Camaros and Firebirds, not to mention most everything else on the road. We even came across a Maserati on the way home. He thought he was a contender, but half throttle in second gear was enough to put a few car lengths between us in a matter of seconds. LS1s? Forget about it. Let's find some turbo 911s.

SOURCES

Comp Cams
3406 Democrat Rd.
Memphis, TN 38118
901/795-2400
www.compcams.com

Crazy Horse Racing
100 Main St.
South Amboy, NJ 08879
732/553-9050

D.S.S. Racing
3550 Stern Ave.
St. Charles, IL 60174
630/587-1169
www.dssracing.com

Holcomb Motorsports
P.O. Box 1473
900 Hardin Rd.
Lumberton, NC 28358
910/739-1699
www.holcombmotorsports.com

MSD Ignition
1490 Henry Brennan Dr.
El Paso, TX 79936
915/855-7123
www.msdignition.com

LaRocca's Performance
1600 Englishtown Rd.
Old Bridge, NJ 08857
732/723-1111

Chapter 5

WINDSOR WORKHORSE

What Makes a 670 Horse 383 Windsor Tick?

By Evan J. Smith
Photos by the Author and Courtesy of the Manufacturers

Over the past 30 years, Ford Motor Company has produced millions of small-block V-8 engines. There were some pretty hot versions throughout the engine's four decades of existence, including the 335hp Boss 351—a Cleveland model with 10.7:1 compression and solid lifters. The '80s brought us the torquey 5.0 H.O., with EFI, roller cam, and plenty of refinement, while the '90s gave us the last, true-blue, high-performance pushrod V-8s. Of course, we're referring to the GT-40ized 302 in the Cobra and the Cobra R's 351.

In the late 90s, Ford had gone modular with the 4.6 and 5.4 powerplants. And while the 4.6 SOHC got off to a rocky start, today's mod mills pack a powerful punch. The Mustang GT cranks out 300 hp, and the king daddy 5.4 DOHC rotates the earth with 550 hp in factory Ford GT trim.

But even with these high-tech overhead cam screamers, the choice of most Ford racers continues to be the small-block. Small-block Ford performance parts still outweigh the modular aftermarket, so it's no surprise the SBF is still the prominent powerplant. Popularity has skyrocketed in the last five years and there are dozens of 302 and 351 stroker kits, along with dozens of cylinder heads, intakes, and other internal parts that are affordable for the average street enthusiast or sportsman drag racer. The 351W wins regularly in NASCAR's Nextel Cup, Busch, and Truck series. And there's hardly a heads-up Mustang drag racing class where a 289, 302, or 351 isn't king. And this continues to push the pushrod technology.

In fact, it has become easy to stuff a hot 302- or 351-based engine in your Ford. And while the 302 is a good choice for your '79–'95 Fox, it's not too hard to add a 351 and a substantial amount of cubes.

The stroked 383 Windsor is built around a Ford Racing Performance Parts M-6061-N351 race block (not shown). These blocks feature four-bolt main caps on the No. 2, 3, and 4 journals and 2.7490-inch (Cleveland) crank-shaft main journals.

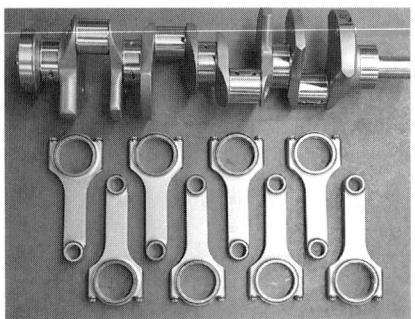

The crankshaft and rods are from Eagle. The crank has a 3.750-inch stroke and the rods measure 6.200 inches.

JE flat-top pistons were used in the buildup. According to Dave Jack of Dave Jack Cylinder Heads, the engine would have benefited greatly from domed pistons and raised compression.

Compare a standard 351 piston (left) with the stroker piston. The stroker piston has a shorter compression height (the distance from the center of the pin boss to the top of the piston) to fit the longer crankshaft stroke inside the confines of the block.

THE NON-MOPAR, NON-CHEVY 383

In the '60s and '70s, Chrysler built many 383 big-block engines, though they never made great race engines. The Bow Tie crowd got 383 crazy in the '80s when they stroked the popular small-block Chevy 350 using a SBC 400 crank with a 3.75-inch stroke. But a 383 Ford? Most think of 331, 347, 393, and 408 as common cubes for a SBF stroker, but a 383? Well, that's what happens when you mix a 351 with 4.030-inch bores and a 3.75-inch stroker crank.

As you know, cubes are an important ingredient when trying to make real-world power (a power adder doesn't hurt either). The factories knew this in the '60s and '70s when big-blocks ruled, but you can't buy a big-block musclecar anymore. So, we began "stroking" small-block engines, and today big-inch small-blocks are in vogue.

Thankfully, all the tools exist for you to "go big," as aftermarket 351W blocks allow you to stretch your small-block engine to 454 (or more) cubic inches. Displacement is your best friend, and when fitted with the right combination of induction and exhaust, the possibilities are endless. Remember, however, that going big means you'll need more cam, more induction flow, and more exhaust flow to maximize performance.

For this exercise, we tagged along with Bob Oster of B&B Performance Machine and Dave Jack of Dave Jack Cylinder Heads as they put the screws to a customer's 351/383-inch stroker. Oster and Jack previously built the 351 into a nice 383, but now the customer wanted more. So an upgrade was planned and put into action.

Oster and Jack originally built the engine using a stock two-bolt 351 block, 0.672 inch lift, 254/258 duration cam, and Edelbrock Victor Jr. heads. With 11.5:1 compression, it made 580 hp at 7,000

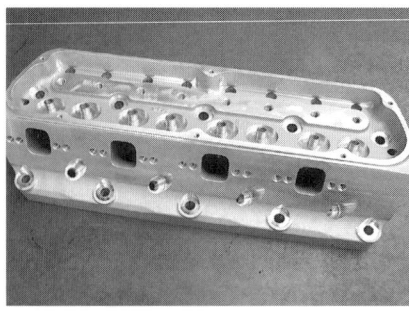

Jack ported a set of FRPP (M-6049-Z304) aluminum cylinder heads for the 383. These aluminum casts are formed from prime A356 aluminum and feature 20-degree valve angles, raised exhaust ports, and manganese-bronze valve guides with semifinished ID. They can be ordered complete or bare.

This is the unported intake port measuring 204 cc. In this trim the heads flow over 270 cfm at 0.500 inch lift. After Jack ports the heads, flow improves to 313 at 0.500 inch lift. Peak flow is 338 cfm at 0.800 inch lift.

Z304 heads come with CNC-machined 63cc combustion chambers. Bare heads require a valve job, while the complete heads can be sent right into action.

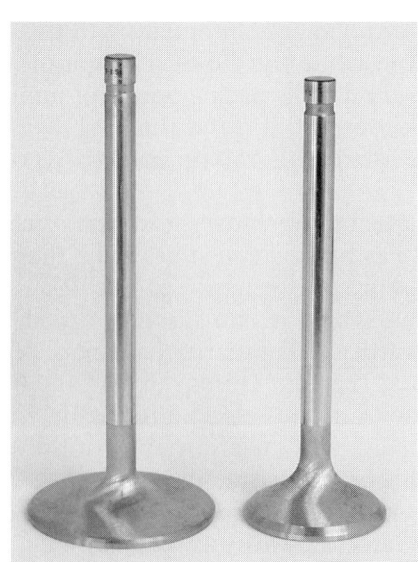

Jack fitted the heads with 2.10/1.60-inch Manley valves.

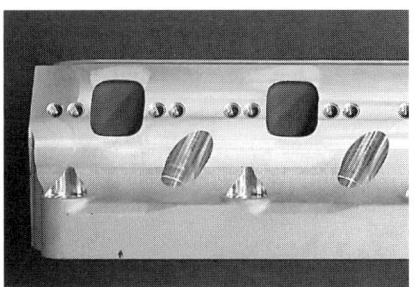

Z304 heads have a thick deck surface and raised exhaust ports. They accept tapered seat and gasket-style 14mm spark plugs.

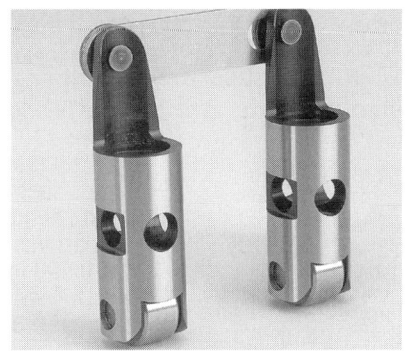

Following the cam are 16 of Crane's Pro-Series mechanical roller lifters, which worked with Manley pushrods and Jesel Pro Series rockers.

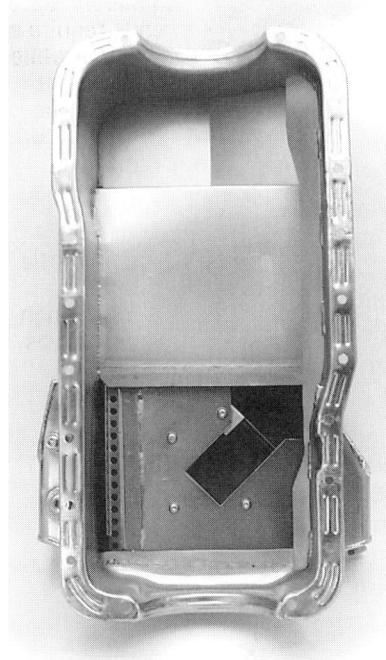

The pan is designed to fit the confines of the Fox chassis. It also features plenty of baffling to keep the oil surrounding the pump pickup.

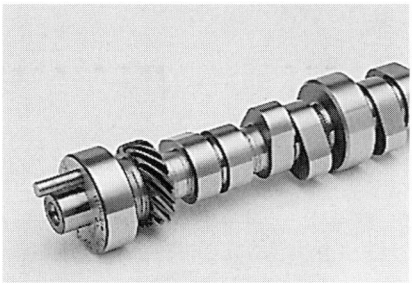

Bob Oster of B&B Performance Machine installed a serious 0.715-inch lift Crane stick with 268/274 degrees duration at 0.050 inch lift.

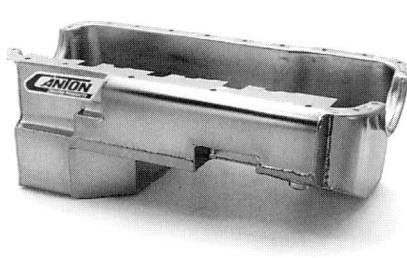

B&B employed a Canton windage tray with a Canton 351 drag race pan with a kickout to contain and control the oil.

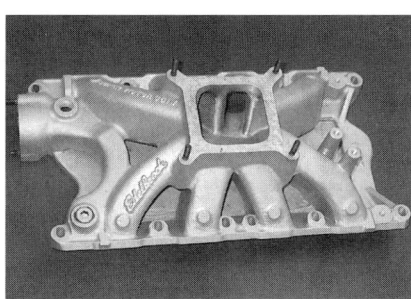

A mildly ported Edelbrock Super Victor sits atop the 383.

Air and fuel is mixed by this Holley 950 HP-series carburetor.

A tapered four-hole spacer from Wilson Manifolds sits under the Holley and adds a few horsepower.

rpm and 500 lb-ft of torque at 5,300. For this go-round, the customer, who is an employee at Downs Ford in Toms River, New Jersey, planned to use a Ford Racing Performance Parts block and the latest FRPP in-line valve heads. It's not that the Edelbrock heads are bad, but he wanted to try the FRPP casts.

It's important to note that the engine used many top-quality parts, but it's not a max-effort race engine. This means it didn't utilize ultrahigh compression ratio, Yates-style heads, a sheetmetal intake, and/or a featherweight rotating assembly. Still, the results were impressive.

THE PARTS

To handle the planned power increase, B&B fortified the bottom end with an FRPP block (PN M-6010-N351) that features four-bolt main caps on the No. 2, 3, and 4 mains, a 9.200-inch deck height, and a Cleveland-sized crank with 2.7490-inch journals. The block was fitted with an Eagle 3.750-inch crank and Eagle 6.200-inch rods, which swing the JE flat-top pistons that came out of the old engine.

"We would have preferred to have maximized this combination with domed pistons and high compression, like 14:1," Jack says, "but we had to work within the customer's budget."

"With [increased] compression," Oster adds, "we could have built over 700 horsepower and even more with some other parts—and that's not too bad for a 383 Ford with in-

The front dress consists of a Jesel beltdrive, a CSR electric water pump, and an ATI Super Damper.

Oster (left) and Jack prepare the 383 to run on the Superflow 901 dynamometer.

Best performance came with the timing set to 33 degrees BTDC and the valve lash adjusted to 0.012/0.012 inch.

The Jesel beltdrive allows quick and accurate adjustment of camshaft timing. It also reduces harmonics.

The engine is fitted with 17/8- to 2-inch headers with 4-inch collectors. It also uses ACCEL wires and MSD distributor. The large velocity stack attached to the carburetor measures airflow into the engine.

The completed 383 is ready to be flogged at WOT.

line valve heads."

While the 383 was apart, Oster installed a slightly larger Crane cam (0.715 inch lift, 268/274 duration at 0.050), along with a Canton oil pan, a Jesel beltdrive, and Jesel rockers.

The short-block was topped with FRPP's aforementioned Z304 small-block cylinder heads. These aluminum casts beat the older "Y" and "X" heads in flow by a bunch, and we've been itching to see them run on an engine. The Z304 heads are offered bare (M-6049-Z304) or complete (M-6049-Z304A)—these came bare

and were prepared by Jack. Prior to installing the heads, he ported the casts using his CNC machine and then hand-finished them using 2.10-/1.60-inch Manley valves and Manley springs (PN 221424-16). This greatly improved the flow, both at peak and on average (see chart for flow data).

After assembly, the 383 was placed on the dyno where it was dialed in. The stroked Windsor pumped out a respectable 645 hp at 7,200 and 502 lb-ft of torque. While overall torque didn't pick up by much, the 383 now makes over 500 lb-ft from 5,600–6,500 rpm, where before it only made 500 lb-ft at one point on the chart.

But the test was not quite done, as the owner had supplied Jack and Oster with a beltdriven vacuum pump from GZ Motorsports. Vacuum pumps are installed on many drag race engines to pull vacuum from the crankcase. This can greatly improve ring seal and oil control, and it can prevent oil contamination.

The GZ kit comes with an adjustable valve to control the amount of vacuum pulled from the engine. A low vacuum draw on the crankcase may not show an increase in power, while excessive vacuum can lead to oiling problems. We followed the instructions and dialed in the recommended 15 inches of vacuum. The result was a gain of 20 hp, bumping our peak to 674 at 7,500 rpm. Meanwhile, torque climbed to 512 lb-ft at 6,100 rpm.

While vacuum pumps have been known to show an improvement, it was shocking to see the 20-horse gain with our own eyes. A series of backup pulls confirmed the results. By the end of the session, we all were impressed with the 383 combination. This is by no means a cheap engine to build, but the value is tremendous. Now, all we have to do is convince the owner to add compression, and he should have over 700 ponies.

FORD RACING Z304 CYLINDER HEAD FLOW FIGURES UNPORTED*

Lift	Intake cfm	Exhaust cfm
0.050	31.1	26.4
0.100	64.6	55.8
0.150	103.0	84.6
0.200	135.3	113.3
0.250	167.0	146.8
0.300	199.0	171.6
0.350	227.2	190.2
0.400	245.2	200.4
0.450	259.8	207.2
0.500	271.9	212.8
0.550	277.3	218.1

*flow figures provided by Ford Racing

CNC PORTED BY DAVE JACK (NO PIPE ON EXHAUST)

Lift	Intake cfm	Exhaust cfm
0.050	37.2	34.0
0.100	74.3	68.1
0.150	111.3	91.7
0.200	148.4	115.3
0.250	184.4	138.9
0.300	220.5	162.5
0.350	248.6	174.7
0.400	276.8	186.9
0.450	295.2	203.7
0.500	313.7	220.5
0.550	320.0	223.1
0.600	326.2	225.8
0.650	330.9	233.2
0.700	335.5	240.5
0.750	337.1	244.3
0.800	338.7	248.0

383-INCH WINDSOR

Tested w/Edelbrock Victor Jr. heads, Crane cam w/0.672-inch lift 254/258 duration at 0.050, 11.5:1 compression

rpm	Torque	Horsepower
4,500	463.2	396.9
4,600	475.7	416.6
4,700	484.7	433.8
4,800	486.2	444.3
4,900	496.7	463.4
5,000	496.2	472.4
5,100	498.3	483.9
5,200	496.3	491.4
5,300	500.8	505.4
5,400	494.8	508.8
5,500	490.9	514.0
5,600	482.9	514.9
5,700	476.0	516.6
5,800	472.6	521.9
5,900	469.1	527.0
6,000	465.1	531.4
6,100	466.7	542.1
6,200	462.3	545.7
6,300	464.4	557.0
6,400	458.9	559.2
6,500	452.9	560.6
6,600	451.6	567.5
6,700	448.7	572.4
6,800	440.3	570.1
6,900	439.8	577.9
7,000	435.4	580.3
7,100	424.4	573.8
7,200	420.5	576.5
7,300	415.6	577.6

383-INCH WINDSOR

Tested w/ported FRPP Z304 heads, Crane cam w/0.715-inch lift, 268/274 duration at 0.050,11.8:1 compression

rpm	Torque	Horsepower
4,500	414.9	355.5
4,600	413.2	361.9
4,700	442.5	396.0
4,800	446.8	408.3
4,900	460.2	429.3
5,000	479.2	456.2
5,100	475.4	461.7
5,200	475.0	470.2
5,300	473.0	477.4
5,400	494.3	508.2
5,500	487.8	510.9
5,600	501.7	534.9
5,700	502.8	545.7
5,800	504.8	557.4
5,900	504.2	566.4
6,000	507.6	579.9
6,100	512.4	595.1
6,200	506.0	597.3
6,300	505.8	606.7
6,400	505.8	616.4
6,500	508.6	629.4
6,600	497.7	625.4
6,700	490.3	625.5
6,800	495.3	641.3
6,900	494.6	649.8
7,000	482.5	643.1
7,100	492.2	665.3
7,200	481.9	660.6
7,300	477.5	663.7

SOURCES

**B&B Performance/
Dave Jack Cylinder Heads**
1056 Randolf Ave.
Rahway, NJ 07065
732/388-1089

Canton Racing Products
232 Branford Rd.
N. Branford, CT 06471
203/481-9460
www.cantonracingproducts.com

Crane Cams
530 Fentress Blvd.
Daytona Beach, FL 32114
386/252-1151
www.cranecams.com

Downs Ford Motorsport
360 Rte. 37 E.
Toms River, NJ 08753
732/349-2240
www.downsford.com

Ford Racing Performance Parts
15021 S. Commerce Dr., Ste. 200
Dearborn, MI 48120
800/367-3788
www.fordracingparts.com

GZ Motorsports
22338 Shake Ridge Rd.
Volcano, CA 95689
209/296-3793
www.gzmotorsports.com

Jesel Valvetrain Innovation
1985 Ceder Bridge Ave.
Lakewood, NJ 08701
732/901-1800
www.jesel.com

Wilson Manifolds
4700 N.E. 11th Ave.
Ft. Lauderdale, FL 33334
954/771-6216
www.wilsonmanifolds.com

photo by Jim Campisano

Chapter 6

A 4.6L NUCLEAR POWERPLANT

**By Michael Galimi
Photos by Tim Stockwell/Fox Lake
Power Products**

For over a decade, the Ford modular engine program has been with us (in both Two-Valve and Four-Valve configurations). In 2005, Ford added a Three-Valve version to the Mustang mix. With regard to its Two-Valve and Four-Valve brethren, the aftermarket has moved forward to overcome many hurdles when it comes to dissecting the Mustang's modular powerplant and making it better.

With the introduction of the '05 Mustang, we were faced with the next evolution of the performance modular engine in the form of a three-valves-per-cylinder setup on the cylinder heads. This was Ford's answer to better performance, increased fuel economy, and lower emissions. The camshaft is controlled electronically through sensors and oil relief valves—further complicating development of performance-enhancing aftermarket parts. The computer advances and retards the camshaft timing. One good thing out of all this is that Ford gave us 300 hp from the factory. That makes this the most powerful base Mustang GT of all time.

While variable cam timing is nothing new to the automotive world, it's weird science to crazy Mustang owners. To make matters more difficult, we have been burdened with a complex computer system to control the variable camshaft timing, among other elements. Would the aftermarket be able to overcome or enhance this new technology? The answer to that question is a resounding "yes." A few companies have already broken down the electronics barrier and have made great strides in tuning the computer system. Alas, as more enthusiasts modify their

The engine has been assembled and is ready to be dropped into Burcham's '05 Mustang. The belts, the blower pipe work, and the rest of the accessories will be bolted in place once the engine is installed. A '97 model Cobra block was used as the foundation for this 600-plus-horsepower engine combination. There were slight differences when compared to the '05 block, but nothing that caused any major problems.

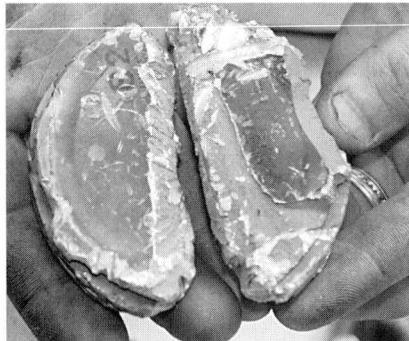

We love carnage! This is what's left of a piston from the stock engine.

Another difference between the earlier-style modular engine blocks and the '05 block is how the engine mounts bolt onto the block. Here is a prototype set of engine mounts from JPC Racing. This set is in the beginning stages of design. The mounts will ultimately be produced and marketed as a way to bolt an early-style block into your '05–up Mustang.

Fox Lake spent a lot of time porting Three-Valve heads to ensure their performance is optimized. The company has a CNC program developed for consistent port work on each set of heads.

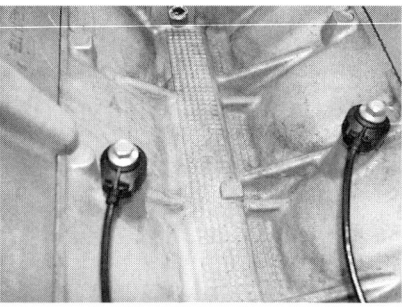

Here's something that's a bit foreign to most Mustang owners—knock sensors. These sensors send a signal back to the computer. If they detect detonation, the computer will compensate by pulling ignition timing. Knock sensors are used in '96–'04 Cobras, but not on the Two-Valve engines in the GT models.

'05-up Mustangs, more parts get broken. But that is when we learn how to fix the 'Stang and make it better.

We hooked up with Justin Burcham of JPC Racing (Glen Burnie, Maryland) to get an inside look at his new engine combination. While most S197 Mustang owners at this point will not be replacing the engine, Burcham's move into this area will lay the groundwork for future projects.

Truth be told, the swap was necessitated after Burcham nuked the stock engine. The Maryland shop owner was on his way to a 9-second run when he missed fourth gear. He got the transmission back into fourth and screamed through the traps at 7,500 rpm. The car did run 10.29 at 136, but the severe horsepower (estimated 700-plus at the flywheel with the nitrous) did not bode well for the stock short-block. One of the connecting rods decided it wanted to see daylight, which effectively made the short-block a pile of junk within a few tenths of a second.

To be honest, we were surprised the stock bullet lasted for so long in a consistent 10-second ride. It

With the intake and exhaust ports optimized, the folks at Fox Lake turned their attention to the valves and valvespring areas. The beefier springs will help the engine rev without going into valve float. They used Ferrea valves for this set of heads.

Final flow numbers check in at 256 cfm (0.500-inch lift) for the intake side. Max flow (269 cfm) occurred at 0.600-inch lift. Stock intake flow was recorded at 220 cfm (0.500-inch lift) and 228 cfm (0.600-inch lift). The exhaust flow was 188 cfm (0.500-inch lift) and 203 cfm (0.600-inch lift) compared to stock, which was 165 cfm (0.500-inch lift) and 175 cfm (0.600-inch lift).

made it through the spring and summer months with weekly trips to the dragstrip and chassis dyno, and let's not forget the street miles logged in that time. The unmolested engine performed quite admirably under such harsh conditions—read 560-plus rear-wheel horsepower with a ProCharger D1SC and a little nitrous to keep things insane. At the time, it was OE from the throttle body to the exhaust manifolds.

The Three-Valve short-block is extremely similar to the modular engines in the '96–'04 models, and the chain setup seems to be a direct carryover, as well. But the camshafts, cylinder heads, and intake manifold are vastly different. The bottom half of the engine was handled by Rich Groh Racing (RGR) Engines.

Groh used a '97 Cobra engine block, as they are supposed to be nearly identical to the Three-Valve engines, though a bit lighter and stronger. They were close enough, albeit there were a few minor issues with the motor mounts. JPC is working on

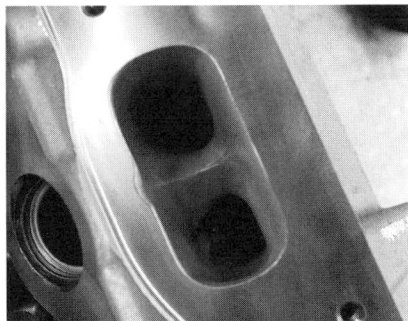

Here's the funky-looking intake port. It may be different, but it gets the job done; that's all that matters to us. There is a straight shot to the combustion chamber.

This is a close-up of the exhaust port.

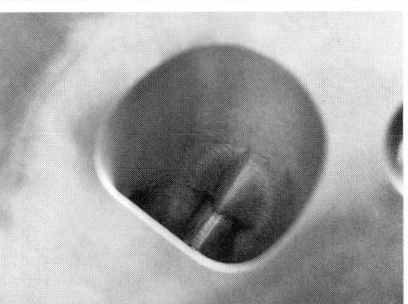

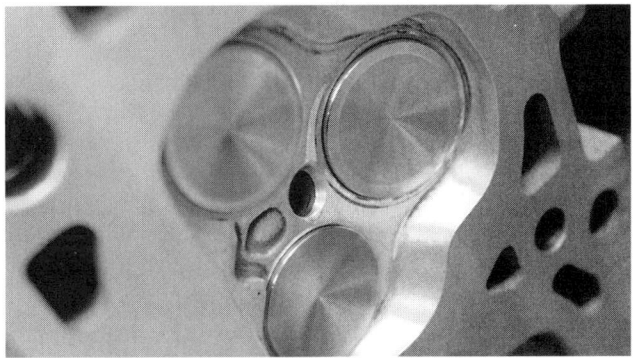

The valves need to drop down 0.350 inch to make it below the deck. That means the intake and exhaust valves are not shrouded by the cylinder wall for most of their cycle. The valves hang below the deck and into the bore for only 0.351 inch to 0.439 inch for the intake and 0.436 inch for the exhaust (maximum lift).

The camshafts mount to the top of the cylinder heads using a system similar to the main caps that hold the crankshaft in the block.

Stock Ford gaskets were chosen for this application. Burcham also uses a set of head studs to hold the heads to the deck.

Burcham torqued down the heads to the block with 80 lb-ft of torque. Be sure to start in the middle and work your way to the outer bolts.

Installing the chains is not as complicated as perceived. The right side must go on first and then you install the left-side chain.

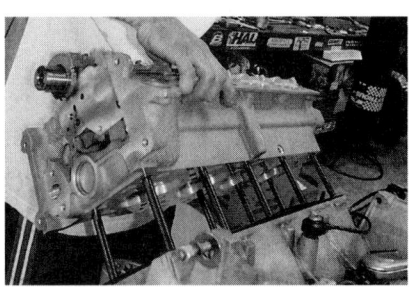

The cylinder heads slide on easily and are ready to get torqued into place.

Here are the components that make variable cam timing a reality. On the left (arrow) is the electronic valve that controls the oil flow. The hub on the front of the cam gear acts as a clutch to advance or retard the cam based on the oil pressure. It is said the camshafts are at full advance and the computer can retard the cam up to 52 degrees.

This is the sensor that communicates with the computer and the electronic oil valve. It mounts to the front cover.

Chain guides keep slack out of the chain. The piece here (left photo) is a spring-loaded follower to keep tension on the chain bracket. These pieces need to be torqued into place (right photo).

The stock intake manifold was reinstalled, and a set of prototype JPC Racing fuel rails were used. These rails will retail for under $300. Burcham also added a set of 75-pound injectors.

Once the chains and the associated guides are installed, bolt on the front cover.

We bolted on an ATI balancer that is SFI-approved. The ProCharger lower blower pulley bolts to this piece. A Mezeire electric water pump was also added along with Metco Motorsport's new pulley kit.

motor mounts to retrofit the earlier modular blocks into your '05-up Stang. Custom CP pistons were chosen for their high quality and forged casting. They are custom-designed for a supercharged/turbocharged combination. The crank was sourced from an '03 Cobra engine and is made of steel. Eight Eagle rods connect the Cobra crank with the CP pistons.

Fox Lake Power Products handled the porting and polishing of the Three-Valve heads. Ron Robart and his crew already have a CNC program specifically designed for the Three-Valve heads. "I think there is more potential in these heads than the Four-Valve heads. The port shape is nice and it's raised up," Robart says. He also mentioned the angle of the ports make it more like a motorcycle head, and the exhaust ports are efficient as well.

The flow numbers may not be what people expect, but Robart wanted to point out the fact that the intake port is extremely efficient. That is

because it has virtually a straight shot into the combustion chamber. Therefore, large flow numbers are not nearly as important in this application. When you have a severe bend in the port, like in the Four-Valve intake ports, you need to get the flow numbers higher. The exhaust may have half the amount of valves as the intake side, but Robart says the exhaust port has a great shape, making it very efficient as well. This could be Ford's best OEM head in terms of performance potential.

Fox Lake installed a set of Ferrea stainless steel valves, which measure the same as the stock pieces. They also upgraded the valvesprings to a double-spring setup. This is probably one of the best benefits because the engine will be able to rev higher and make more horsepower.

Unfortunately, there were a few components we would have loved to change, but at the time, nothing was available. The camshafts were left stock, but Comp Cams will soon be releasing several different grinds for these engines.

The other component we left stock was the intake manifold. We are not sure of the flow numbers, but the intake definitely looks a lot better than the OE Two-Valve piece. The runners are shorter to promote better breathing at higher rpm. The stock Two-Valve intake had to help create low-end torque because of the 4.6's small displacement. One would also think the Two-Valve engine would be a rev-monster, but that isn't the case because the manifold is restricting it. The Three-Valve engine may have the same 281 cubes, but the variable cam timing made a huge difference by creating a broader torque curve when compared to its Two-Valve cousin.

Ford did its homework, and this engine is more than capable of pulling serious rpm with the correct valvesprings in place. Burcham has turned 7,500-

The throttle body remained stock, and Burcham wanted to say thanks to Jessie Kershaw and Jeff Dunne of Ford Racing for getting him a new intake manifold and throttle body in a hurry.

Burcham bolts on the ProCharger D1SC head unit. The supercharger system features a large air-to-air intercooler and can supply up to 32 psi of boost if Burcham chooses to bump up the impeller speed. Of course, he is going to keep this engine combo sane and restrict himself to "only" 22 pounds.

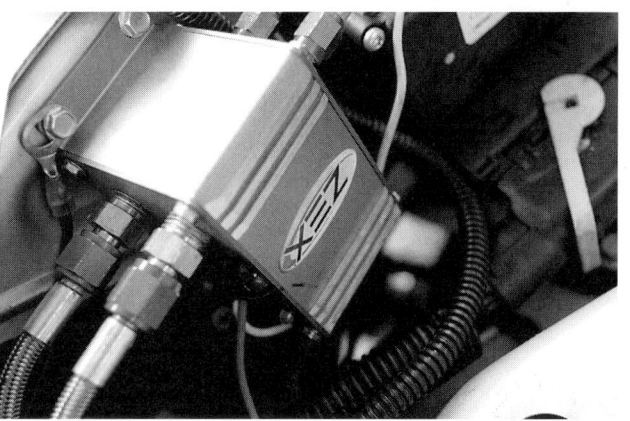

An SVT Focus fuel pump is piggy-backed to the stock in-tank pump and is enhanced using MSD's fuel pump booster box. This piece increases voltage to the fuel pump, thus pushing more volume through the system. A –8 line was used to feed the fuel rails.

We installed a Zex nitrous control box. Burcham saves the nitrous for the "big races."

plus rpm (without valve float), and the intake manifold seemed to allow the engine to breathe deep enough to achieve it. He did report that optimum power with his new engine is around 6,900–7,000 rpm. This is with the stock camshafts in place, too. "I tried shifting 7,000 and higher with the stock engine, but the valvesprings were too weak and the engine would float the valves," Burcham says. Fox Lake fixed that problem with a stiffer set of springs designed for high-performance usage.

As a side note, we wanted to mention some observations in testing. Burcham spent some time at ProCharger with this car and found that the more boost stuffed into this engine, the lower the rpm of peak power. For example, 15 psi (565 rwhp) made peak power at 6,800 rpm, but 17 psi (585 rwhp) had a maximum horsepower rating at 6,500 rpm. At 20 psi (615 rwhp) of boost, the

peak power rpm occurred at 6,200. This was with the stock engine, not the built engine with the better heads and valvesprings. Burcham attributed the rpm variances to the weak OE valvesprings not working properly with the large amount of boost (or cylinder pressure).

We would certainly like to do a cam swap in the future to see what the Comp Cams stuff is worth over the stock sticks, but that will have to wait. For now, we'll just reinstall the stock camshafts, even though it's something we know has to be limiting performance. After all, Burcham's engine does get stuffed with a lot of boost from a ProCharger D1SC blower, and it sometimes gets a whiff of nitrous when the big money is on the line.

With the engine together and the factory electronics still in place, it came time to tune the beast. Mike Carlson of ProCharger provided assistance in getting the tune-up correct. He worked with Burcham via phone and Internet to get the programs loaded into the car through a Predator handheld tuner. After sneaking up on the proper tune-up using an air/fuel meter, the car spun the Dynojet chassis to the tune of 640 hp and 544 lb-ft of torque (at the tires).

The car has also received numerous upgrades to

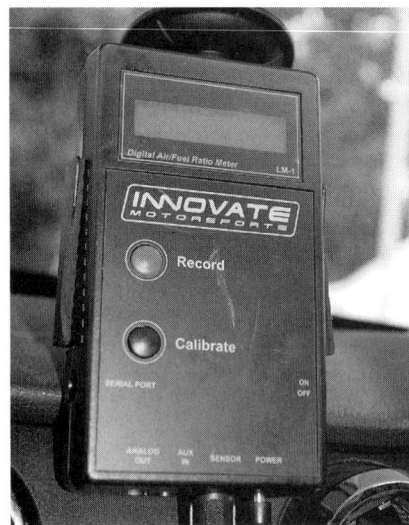

An Innovate Motorsports air/fuel meter is mounted to the windshield with a suction cup. It keeps the meter in plain view so Burcham can keep tabs on his engine with ease.

the driveline to handle the extra horsepower. First on the list was a "real" transmission designed to take supercharged abuse. With assistance from a Swarr Automotive conversion kit, Burcham installed a Tremec TKO 600 five-speed transmission, an SFI-approved bellhousing, and a Centerforce clutch. A new JPC Racing one-piece driveshaft was also installed. The one-piece driveshaft saved 30 pounds in rotating weight and is stronger than the stock two-piece unit.

We're confident the horsepower is there to achieve a nine-second run in his full-weight street car—complete with air conditioning, stereo, and other amenities for life on the streets.

SOURCES

ATI ProCharger
14801 W. 114th St.
Lenexa, KS 66215
913/338-2886
www.procharger.com

Fox Lake Power Products
6060 Dalton Fox Lake Rd.
N. Lawrence, OH 44666
330/682-8800
www.foxlakeracing.com

Justin's Performance Center/JPC Racing
217 Thelma Ave.
Glen Burnie, MD 21061
866/572-7223
www.jpcracing.com

Metco Motorsports
109 N. Park Dr.
Anderson, SC 29625
864/332-5929
www.metcomotorsports.com

A large air-to-air intercooler is standard in the ProCharger kit. It mounts comfortably up front, behind the bumper. The lower valance opening provides ample airflow over the fins to chill the boost inside the intercooler.

The engine is ready to get dropped into Burcham's street/strip '05 Mustang.

Chapter 7

A NICE 4.9L STROKER

We Prepare to Chill Out the Competition With a 4.9L Modular Stroker from Coast High Performance

By Evan J. Smith
Photos by Evan J. Smith and Richard Holdener

PART I

When we laid out the plans for our '01 GT project car, we never expected to run low 11s at 126 mph using many of the stock engine parts, namely the factory bottom end. But thanks to a careful selection of induction components, smart tuning and clean powershifts, the GT runs hard and the 4.6 has lived a healthy and rather potent life. In its three years as a test mule, we've used our pony as a road racer, drag racer, and as a daily driver that sees about 70 miles a day rolled onto the odometer. Ice Box, as we call it, has also been a test bed for many new Mustang parts, from suspension to superchargers and even the latest in body wear.

Like many enthusiasts, we've been curious about how much abuse the stock 4.6 can take. To date, our engine, which wears a Vortech SQ1 blower, Patriot Performance heads, Comp Stage 2 cams, a ported Bullitt intake, JDM Engineering headers and catalytic H-pipe and SLP cat-back, has pumped out 540 rwhp and 470 lb-ft of torque, also measured at the wheels. It's run 11.20s in the quarter at 126 mph and knocks down over 15 mpg in the city/highway loop while maintaining full emissions legality! I call that well-rounded, but it's not perfect. Not yet, anyway.

The truth is, we want 10s. And while we could simply turn up the wick (read: boost and rpm) on the current combo, that's not what we've decided to do. Yeah, yeah, we know full well that we could push what we have a bit harder and it would probably produce a few 10-second time slips.

Our Coast High Performance stroker is up on the stand and ready to be dropped into Project Ice Box, our '01 Mustang GT. Once installed we'll turn up the wick and shoot for maximum boost and power.

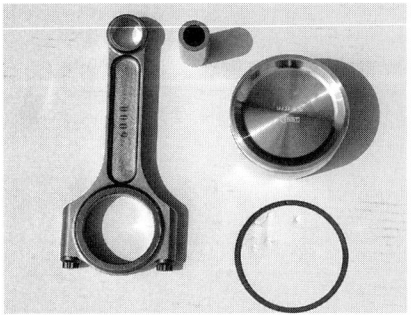

The CHP stroker engines come equipped with Probe pistons (ranging from .020- to .070-inch over) depending on application, along with Probe connecting rods and 8620 chrome-moly pins. For extreme applications (10-psi of boost or more), CHP offers Tool Steel wrist pins that cost about $100 extra. In deference to our supercharged application, we went with the .020-over, 300-cube Pro Street short-block, not the 310.

Probe forged pistons feature an offset pin placement and are fitted with a tight piston-to-wall clearance. For our application we chose dished pistons with a 21.1cc volume. This should give us about 9.1:1 compression.

Our CHP stroker features a 3.750-inch crank and .020-inch-over pistons. This calculates to 300.4 cubic inches. With the .070-inch pistons, CHP can make your Mod motor as big as 5.1L or 310 cubic inches.

CHP offers two crankshafts for its Modular strokers. Street Fighter engines get the cast offset-ground cranks, while the Pro Street kits are equipped with a forged-steel version. Both models feature a 3.750-inch stroke that is .200-inch longer than stock.

The Probe connecting rods are of the I-Beam design and are made from 4130 chrome moly. They are bushed on the small end and balanced along with the pistons and the crankshaft.

Before putting the engine on the stand we had to tap in the rear cover alignment pins.

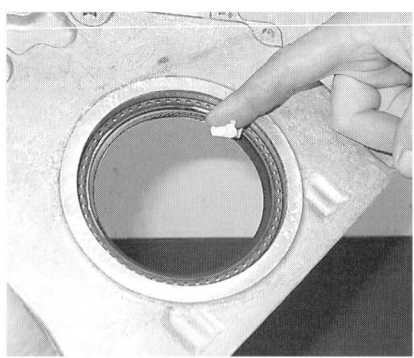

Then we covered the rear main seal with a liberal amount of white grease.

Next, we applied silicone to the back of the cover...

...and the cover was installed.

But while we'd probably get in the 10s, that route is a surefire way to ventilate the block, ending our fun.

In our opinion, we've pushed the stock short-block just about as far as it should go. Sure, some have made more power, some have gone quicker with similar parts—but others haven't. We've seen 'em blow and it ain't pretty. So, rather than chance knocking the rods out the side of the block, we're going to yank the engine and slip in a better short-block with more cubes and better internal parts.

Enter Coast High Performance and its line of Modular stroker kits. CHP is well known for its Street Fighter 347 small-block strokers and now the California-based firm is offering the same quality and power potential in the Modular line of engines.

Stroking, or to be more technically correct, upstroking, has long been one of the many tricks used by hot rodders to increase the displacement and, in most cases, the power of an engine. This essentially means you can use the same (stock) block and get more cubes, without having to swap in a larger "big-block" type engine. In other words, rather than dropping in the physically larger 5.4, you can simply make your 4.6 internally larger by stroking it.

Modular engines feature a second rear seal that must be installed once the cover is in place. Shaun Lacko of JDM Engineering stated that the second seal is prone to popping off during installation, so he devised this trick of using two flywheel bolts and two washers to hold the bottom of the seal on the crank while he presses the top into place. With the rear seals and cover installed, the engine is now ready to go up on the stand.

The rock-solid CHP bottom-end should be able to handle just about anything we can throw at it.

According to Chris Huff of CHP, the 4.6 engines are limited in the stroke department because of the length of the cylinder bores. At BDC the pistons stick out of the bores, but this is not a problem with the CHP kit because of the careful engineering of the parts.

In order to achieve the goal of increased displacement through upstroking, you'll need to install a different crankshaft with the connecting rod journals moved further away from the crank centerline. This increases the length or distance that the pistons travel in the cylinder, however, a new set of connecting rods and pistons must be used to physically fit the package inside the engine. Otherwise, the increase in stroke would cause the pistons to slam into the heads on the upstroke and the pistons would come out of the bottom of the bores on the downstroke. In addition, there can be clearance issues between the camshaft and the rods (in a pushrod engine), and between the rods or rod bolts and the block. So the kit must be designed well and installed carefully.

The CHP Modular kits utilize crankshafts featuring a 3.750-inch stroke, which is .200-inch longer than the stroke on the stock 4.6. While this doesn't sound like much, the increase alone adds about 9 cubes to the 281-inch Mod engine. To take the package further, CHP bores the blocks from .020 to .070 inch over the standard bore size. An overbore of .020 inch increases the displacement to 300, while going out to .070 over makes a 310.

"We start out with new 3.552-inch [bore] stock blocks," says Chris Huff of CHP. "We can go as big as 3.625 with the bore, which is .073-inch

over, but we like to stick with about .020-inch over in high-boost applications." And since we plan on turning up the boost in project Ice Box, we went with a .020-inch overbore. This gives us 300-cubic-inches or 4.9-liters of displacement.

"We have limited the stroke, because at BDC the piston is coming out of the bore too far," stated Huff. "We have still kept more than a 1/2-inch of [piston] skirt in the bore and we offset the piston pin to reduce rod angle. The larger the stroke the more dramatic the rod angle and that creates more side load on the piston, along with more noise and reduced engine life. We help solve this by offsetting the pins in the pistons."

The CHP strokers use crankshafts with a 3.750-inch stroke and small 2-inch rod journals. Don't be alarmed by the small journal size; they are plenty strong and they greatly reduce rotating weight. In addition, the engines get Probe pistons with a special compression height and Probe 5.950-inch rods, which are slightly longer than the stock rods.

"The [Modular] engine is such a tight package from Ford that we had to juggle the numbers to make it [the stroker kit] fit," stated Huff. "We had to put the pin in a place where it intersects the oil ring, but we have a skirt that allows a tight piston-to-wall clearance, so there will be slightly more oil consumption than stock, however, it's nowhere near excessive."

It's one of life's little sacrifices, but to make power you have to have cubes and stroking the 4.6 produces torque, which this engine needs badly.

Currently, Coast High Performance offers a few versions of this kit. The least expensive is the Street Fighter 5.1 with a cast "offset-ground" crankshaft, which retails for $1,999 as a kit, or $3,499 assembled with a new block. The Pro Street kit comes with an aftermarket forged-steel crank and sells for $2,599 (in kit form) and $3,999 assembled with a new block. We opted for the Pro Street version and we added the aluminum main stud

Our 300-incher is itching to make some power. Next we'll get Ice Box in the shop and begin to do the swap.

We got to the business of preparing the Coast High Performance short-block by slipping on the new ROL head gaskets.

girdle option. In addition, CHP offers Tool Steel wrist pins for an additional $100, a small price to pay for much stronger pins. Chris Huff recommends the Tool Steel option in any engine that will see more than 10-psi of boost.

From here our plan is to adapt the CHP Pro Street short-block to our already potent combination of heads, cam and intake. At the same time we'll also swap our Vortech SQ-Trim, for a hotter T-Trim. This combination will provide us with a bulletproof bottom end and a free-breathing top end. The quality of the rotating assembly will also allow us to spin the beast past the stock rev limiter without fear of the worst.

In previous dyno tests, we saw power still climbing when the engine tagged the 6,250-rpm factory rev limiter. We were afraid of revving it too high, so we didn't raise the limiter, though through the magic of computers we could have. Now, we'll be able to go to 7,000 and beyond—if we deem it necessary.

PART II

In order to keep up with the slew of amazing Cobras roaming the northeast, we decided that more cubic inches and boost was in order. We contacted Coast High Performance, Vortech Engineering, and JDM Engineering to help out, and they had all the answers. Utilizing the stock short-block with ported heads with larger valves and Comp Cams Stage 2 bumpsticks, Ice Box had produced 540 hp (at 6,250 rpm) and turned a best quarter-mile time of 11.26 seconds at 126 mph. That's pretty sporty for a 3,600-lb., daily-driven, emissions-legal GT, but as I stated in the opening, it was time to find more. More being 10s.

But in order to find 10s, more also meant we'd need to up displacement and boost, so CHP sent over a complete 300-cube short-block stuffed with a stroker crank, Probe pistons and rods, and a main stud girdle, just for good measure. Meanwhile,

Shaun Lacko of JDM Engineering then popped our Patriot Performance cylinder heads into place.

Vortech swapped out our SQ supercharger unit for a more potent T-Trim. Putting the puzzle together and tuning this monster were the fine folks at JDM Engineering in Freehold, New Jersey.

It's important to note that our machine was running in top form when we tore it apart. There was no death smoke, no detonation, and no unusual noises. With that, we could have turned up the boost on the hopped-up stock short-block combination and pushed it further. It's likely to have gone much quicker, but for how long? We realized that making much more than 540 rwhp with the stock crank, rods, and pistons was not a good idea. At *MM&FF* we're into making power, but we don't see the point of a one-shot wonder. We're big on durability and longevity, so out with the old and in with the new.

Speaking of new, our new Ford Modular block was prepped by CHP with a steel up-stroked Cobra crank, Probe rods and pistons, and a Probe main stud girdle. CHP can actually open the cylinder bores up as much as .070 inches (to make 5.1 liters/310 cubes), but because of the high boost

Lacko torqued the heads to the block. The stock hardware required a special torquing sequence that goes as follows: 30 ft-lb of torque, then turn 90 degrees, then loosen each bolt at least one full turn, then tighten to 30 ft-lb, turn 90 degrees, and then turn 90 degrees once more. This ensures the proper stretch on the head bolts and a good seal from the gasket.

The old timing chain setup was replaced and the front cover was installed.

Other than the short-block, we were using all of the parts that came off the original engine. Here, the cam covers were bolted to the heads.

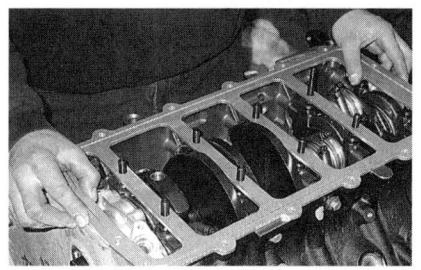

CHP supplied us with this lower main girdle, which slipped neatly over the main studs. Unlike a Windsor or 289/302, where the girdle sits completely inside the oil pan, the Modular main girdle fits between the block and the oil pan; therefore, it is necessary to use Permatex Ultra Black RTV sealer (or similar) on either side of the girdle.

With the heads, timing cover, and cam covers installed, we rolled the engine over so we could prepare the bottom end.

The underside of the girdle is supported by jam nuts that get run up the girdle and pre-loaded with 1/6 of a turn. Then the girdle is torqued to 20 ft-lb.

Lacko applied a coat of Permatex sealer to the girdle and he slipped on the oil pan. It was getting late, so we called it a night.

levels we plan on, we decided to go with a .020-inch overbore to keep some extra rigidity (this was at Coast's recommendation). When combined with the 3.750-inch stroke, our .020-inch over short-block arrives at an even 300 ci, just about 20 more than stock. And to that we'll add our Patriot-ported PI heads, along with the rest of Ice Box's components, which includes the aforementioned Vortech blower, a ported Bullitt intake, JDM headers, and SLP Loud-Mouth exhaust. Then, once the beast is running, we'll add an Anderson Ford Powerpipe, which is sure to unlock even more power.

To date, we've dressed the CHP short-block and slipped it back between the rails. We also took the opportunity to swap out the 37,000-mile factory clutch with a RAM Powergrip Performance series clutch supplied to us by Downs Ford Motorsport in Toms River, New Jersey. The RAM clutch features a blueprinted high-clamp pressure plate, a 900/300 series clutch disc, a new throw-out bearing, and a clutch alignment tool. We also installed a new RAM flywheel from Downs.

Like the stock short-block, the Ford clutch was doing a great job, but with the potential for about another 100 hp, we felt it was time to upgrade. Joe Amato, the top dog at Downs Ford Motorsport, sells a lot of different clutches for different applications. When we told him about our combination, the car's daily-driver status, and the horsepower level, Amato had no reservations about the Powergrip clutch. The RAM unit should provide us with plenty of grip, but without massive pedal pressure. We've yet to use this style RAM unit so we won't know how it works until we get to the track.

Though the motor and clutch

When we arrived the following day, Lacko had installed the ported Bullitt intake along with the front dress and the new Vortech T-Trim blower.

The engine was hooked to a cherry picker and lowered into the waiting engine bay.

Before dropping in the radiator, we aligned the drivebelt.

The T-trim is designed to produce more boost than the SQ, which should help get more air into those extra 20 ci.

The Mod missile was bolted down and we began the task of connecting the bevy of vacuum lines and wire connections.

Once the engine was in place, we attached the new RAM Powergrip Performance clutch.

One side of the RAM 900/300 disc uses an organic facing material bonded to a steel backing plate. This disc also uses a marcel for smoother application and reduced chatter.

The opposite side of the disc uses more aggressive sintered iron (powered metal) pads as friction material. The sintered iron has a very high coefficient of friction, which gives it tremendous holding power without relying on sheer clamping force. This equates to a softer pedal feel, yet with lots of overall grip.

Team JDM attached the new flywheel and clutch assembly.

Last, the transmission was lifted into place and bolted to the back of the engine. Next month we'll have the exhaust hooked up and the Ice Box will do battle with the dyno rollers and the quarter-mile.

are in, we still have a bit of work to do. Ice Box is in running order and D'Amore is finalizing its tune-up. Initial reports are we'll have around 700 ponies at our disposal. Wow.

PART III

If we've learned anything about hot-rodding in the last 15 years, it's that you should always expect the unusual and unthinkable. Sometimes parts go on without a hitch, other times they don't. This was one of those times when even the best-laid plan needed modifying.

As it turned out, changing the short-block and blower was not as simple as we originally planned. We encountered a few bumps in the road—thankfully, we had a good team behind us. Oh, the engine went right in and purred like a kitten, but we battled belt slippage and a lean condition. After a few long nights, we sorted it out, and we're glad to report our Oxford White GT is up and running—and doing quite well, we might add.

The Tech Of It

With Ice Box operational, Jim D'Amore put a few test miles on the engine and clutch, loaded the GT on the Dynojet, and the tuning began. After establishing that the engine was in good running order, D'Amore went to work using Superchips Custom Tuning software. He was concerned with running the engine lean, so the initial pulls were limited to 5,000 rpm.

Thanks to the combination of parts from Anderson Ford, Coast High Performance, Vortech, JDM Engineering, and a few others, our '01 GT is a real screamer, producing well over 700 hp at the crank and 620 hp at the wheels. Based on our e.t., trap speed, and weight, we'd estimate about 750 crank horsepower!

The Anderson Ford Power Pipe was worth 3 pounds of boost, which translates into about 40 hp. The Power Pipe is huge. It increases the airflow into the engine when compared to the inlet supplied in the original blower kit.

The Ice Box gets strapped to the dyno and the air/fuel meter is connected to the exhaust.

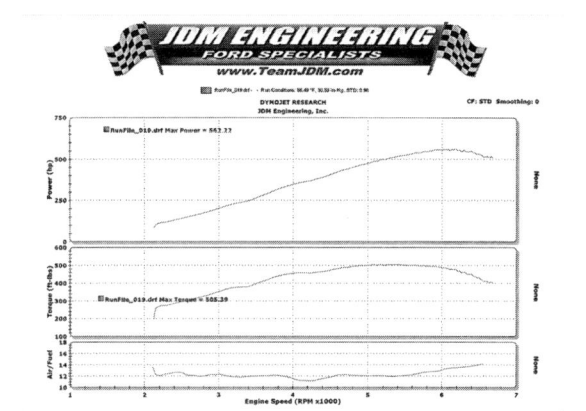

A lack of fuel flow caused a lean condition, which limited the horsepower our new 300-inch CHP stroker could produce. We solved this by upgrading the fuel system in Project Ice Box.

Right off the bat the engine showed signs of greatness, but on the first full pull D'Amore noticed a problem: "Above 4,500 rpm, this thing really comes on strong, but it flattened out just before 6,000 rpm because the belt was slipping like crazy."

The problem, according to the folks at JDM, was belt stretch. This occurred from use and mileage, so a new, slightly smaller belt was installed. With the belt tight, D'Amore flexed the muscles on the GT, and a second problem popped up. The blower was pumping serious boost—15 psi, to be exact—but D'Amore noticed a lean condition and shut it down. The Dynojet's data showed the air/fuel ratio climbed to 13:1 at 6,500 rpm, but it began going lean just before 6,000. Despite the lack of fuel at high rpm, the Coast High Performance 300 made 560 hp and 505 lb-ft of torque.

D'Amore commanded more fuel through the SCT software, but the existing fuel system couldn't answer the call. The engine remained lean, so more mods were in order. The answer was to modify the fuel system to provide more flow; JDM had the answer.

With the old combo, our GT was outfitted with twin '03/'04 Cobra pumps, which replaced the single GT pump. The Cobra pumps were tied together to feed one single outlet line (in the tank), which then fed the stock fuel rail and 42-psi injectors.

To increase the flow, JDM decided to separate the lines coming off the pumps to get more volume

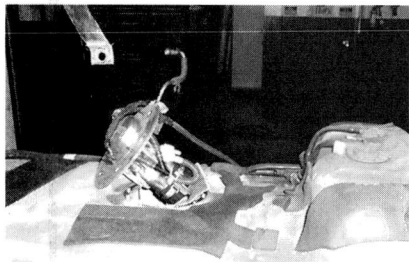

We began by removing the fuel tank, then the twin Cobra pumps.

Lacko drilled the top of the hat and welded two new fittings in place. You can see how each pump feeds each of the fittings.

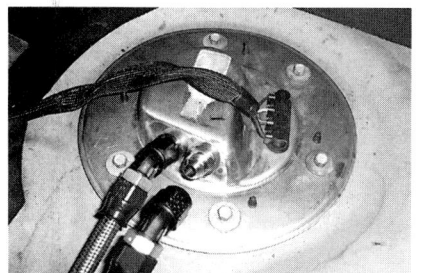

The pumps were placed back in the tank and the new lines were attached.

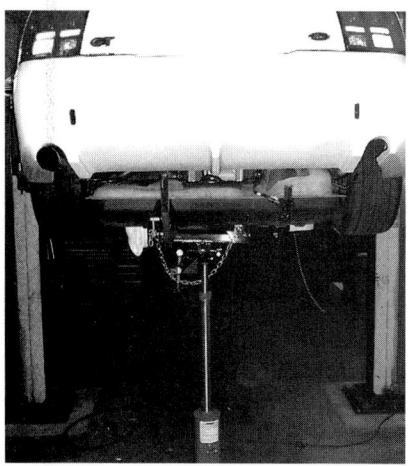

Lacko and crew reinstalled the tank in Project Ice Box.

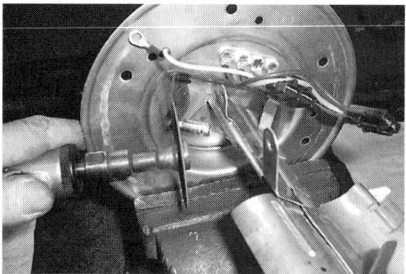

Here, Shaun Lacko cuts off the single outlet from the top of the fuel-pump mount. It will be reworked to have twin outlets.

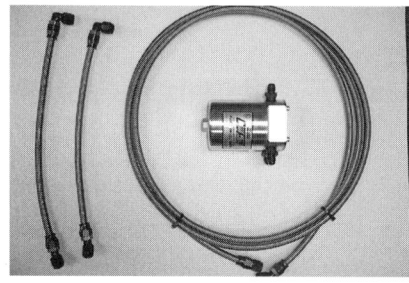

JDM Engineering designed this new system that uses –6 lines and a Mallory fuel filter.

The new Mallory filter was mounted above the rear housing and the lines were connected.

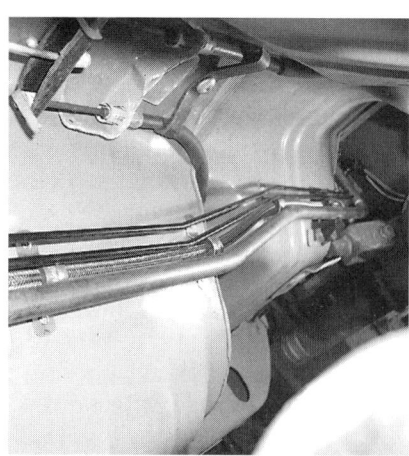

The new fuel line was run up to the engine.

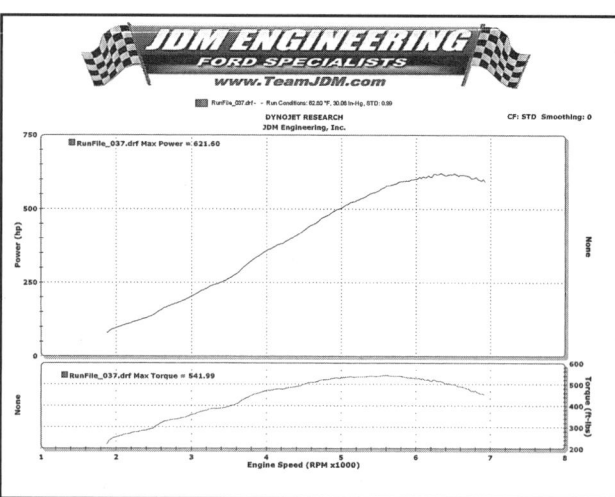

With the adequate fuel flow, D'Amore was able to get back to the business of making power. The dyno graph shows that the 300-incher is making 621 hp at the rear wheels and 541.99 lb-ft of torque. This equates to an estimated 730 flywheel horsepower and an amazing 2.44 hp per cubic inch.

to the main feed line, which would also be replaced, along with the injectors. Lacko removed the tank from the vehicle and dismantled the Cobra pump assembly in order to separate the pumps. Now, each pump feeds into an external fuel bulkhead that's attached to an external Mallory canister-style fuel filter. The filter has two inlets and a –6 line feeding the engine. The stock rails were maintained, but 60-pound Mototron injectors were installed.

This increased fuel flow to the engine and enabled D'Amore to get back after the tune. When he did, the results were outstanding. Horsepower skyrocketed from 562 (the number achieved prior to the fuel-system upgrade) to 621 at 6,400 rpm. Meanwhile, torque climbed to 541 lb-ft at 5,600 rpm. It's important to note that while the GT was making roughly 80 more horsepower than before, D'Amore explained that the huge airflow increase caused the SCT Big Air 90mm mass air meter to max out. So he was unable to unlock the full potential of the engine. In other words, once he installs a differently calibrated mass air meter, he feels there's another 30–50 hp to be had. But that will have to wait for another day.

Nevertheless, we were glad to see the new combination was producing more than 500 hp from 5,200 to redline and over 500 lb-ft of torque from 4,500 to 6,400 rpm. But how would this translate into track numbers?

While dyno numbers are great, we wanted to enjoy the power and find out how much improvement there would be at the strip. Our first runs were made on Nitto Extreme Drag 18-inch tires—the same ones employed on our 22-mile-one-way daily commute.

After dropping the pressure to 20 psi and heating the Nittos, I was able to nail a smooth launch and the GT took off like a missile. Power shifting the Pro 5.0 Power Tower was a breeze...

...and Team MM&FF was rewarded with a 10-second timeslip in 100 percent street trim—in spite of a road-race suspension that limited our 60-foot times to a relatively unimpressive 1.80.

After each run, we cooled the engine using ice and electric fans.

11th-Hour Track Thrash

With the car complete and the dyno numbers in the bag, there was but one thing left to do—drag test. Our man D'Amore would have loved one more day to tune, but there just wasn't time. In fact, Raceway Park had officially closed the evening before, but, thankfully, we managed to coax the RP staff into leaving the power on for one more day. Hey, we love the chassis dyno, but you want real-world dragstrip numbers. So here we go.

For the first few runs we decided to leave the car in street trim, meaning we'd keep the Nitto Extreme Drag tires, along with the Eibach front and rear antiroll bars, in place. With so much power on tap, we launched easy on the first hit and scored an 11.48, but at a whopping 131 mph. A more aggressive launch netted an 11.26 on the second hit and, after a cooldown, the third run produced an 11.05. Amazingly, the tires hooked like they had teeth, and we were darn near the 10s on the drag radials. Although we planned to go to the stickier M/Ts, we were determined to run in the 10s with the Nittos.

Thankfully, we nailed the next launch and it happened—a 10.93 at 133 mph. The level of power was just insane, especially for a Two-Valve GT. Heck, 133 mph is up in the territory of a Cobra when equipped with a Kenne Bell or a Whipple. Geez, 133 is more than you'll see in a $600,000 Ferrari Enzo! As D'Amore stated, this combination is animal-like above 4,500 rpm. It pulled and pulled, then pulled some more. We shifted right at 6,700, but could have taken it higher. It never stopped pulling.

With a 10-second drag-radial run in the books, we attached the 26x11.5/17-inch M/T ET Street tires. With the new combo and sticky Mickey tires, Ice Box ripped through the E-Town quarter-mile in 10.701 seconds at a whopping 132.78 mph. Equally impressive was that our GT ran an "off-the-street" 10.90 at 133-mph run with radial tires,

While the Nitto tires provided excellent bite, we wanted to try a set of Mickey Thompson ET Streets. The car actually hooked better on the M/Ts, but the extra height killed a bit of the gear and we couldn't realize the full potential of the extra traction. The 60-foot time improved from a 1.807 best with the Nittos to a 1.653 with the ET Streets. If we were going to continue using the M/Ts, we'd switch back to 4.10 gears.

Run	60-foot	ET/mph
1	1.988	11.480/131.56*
2	1.836	11.267/123.68*
3	1.886	11.051/133.24*
4	1.807	10.939/133.29*
5	1.672	11.026/130.62**
6	1.653	10.705/132.78**

*With Nitto NT555-R Extreme Drag
285/35ZR/18-inch tires
**With M/T ET Street 26x11.50/17-inch tires

catalytic converters, fat front tires, and both antiroll bars in place. It weighed 3,630 pounds and we drove it home after the test.

As do many projects, this one proved to be time-consuming and extremely challenging. We were pitted with some serious deadlines and there were no second chances, but we completed the mission and brought home the goods. We couldn't have done it without the crew at JDM. We'd like to thank them for the late hours, along with the staff of Englishtown's Raceway Park, who kept the electricity and water turned on for a few extra hours so we could complete the test in time to make our deadlines.

SOURCES

Anderson Ford Motorsport
1001 State Rte. 10 W.
Clinton, IL 61727
217/935-2384
www.andersonfordmotorsport.com

Coast High Performance
2555 W. 237th St.
Torrance, CA 90505
310/784-1010
www.coasthigh.com

JDM Engineering
60 Jerseyville Ave.
Freehold, NJ 07728
732/780-0770
www.teamjdm.com

Downs Ford Motorsport
360 Rte. 37 E.
Toms River, NJ 08753
732/349-2240
www.downsford.com

Ram Clutches
201 Business Park Blvd.
Columbia, SC 29203
803/788-6034
www.ramclutches.com

ROL Gaskets
3100 Camp Rd.
Oviedo, FL 32765
800/327-1027
www.rolmfg.com

Vortech Engineering
1650 Pacific Ave.
Channel Islands, CA 93033
805/247-0226
www.vortechengineering.com

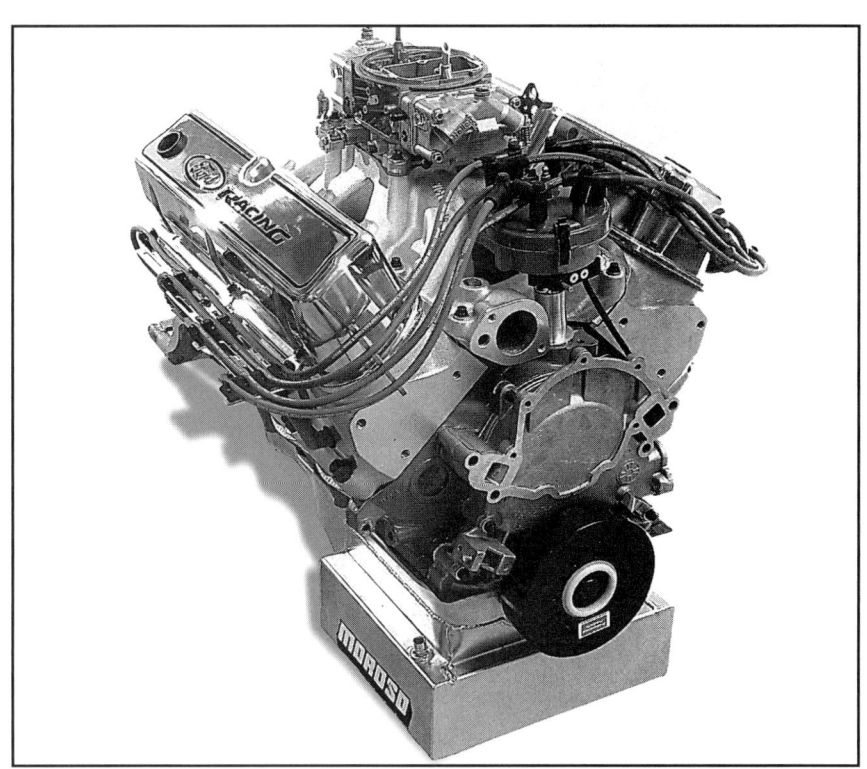

Here is our 370-incher ready to be loaded on the dyno. It's minus the water pump because the pump used is one specifically set up for T&L's dyno.

Chapter 8

5.0 to 6.0 STRETCH

A Stroker 347 Used to Be the Hot Ticket, But this Pumped Up 370 Makes 606 HP on Pump Gas

Text and Photos by David Vizard

OK, the 5.0 to 6.0 stretch, is, admittedly, a little short of the truth. It makes a catchy title, but the reality is, we are actually looking at a 1.124-liter stretch (1,124 cc) and as we know, more cubes are better than fewer cubes. This makes our supposed 6.0 small-block nearer to 6.1. All this in a block that still sports an 8.2 deck height.

Knowing that cubes are king, and with the ease of extracting extra cubes from the taller 351 Windsor block, you may ask why stick to the shorter deck height? Well, there are several reasons. First, although it's not the big issue it used to be, there's just more room for headers between the suspension towers with the 8.2-deck block. Second, you get a lower center of gravity, which is a plus if fast cornering is of any interest. Third, there is a stealth factor, and fourth, a 351 Windsor weighs about 90 pounds more than the thin-wall cast 5.0 block.

BIG-BORE BEEF

Before we go further, let's visit our goal, which is simply a quest for cubes. Then to best utilize those extra cubes in terms of cam specs, heads, intake, and so on, and to keep an eye on how much the extra cubes will cost. We have successfully been to 347 inches with a 3.4-inch stroke Scat crank in a stock block. Still working with Busch Engine builder Lloyd McCleary (T&L Engines) for all our machining and assembly, our first move was to make a decision as to what block to use. Here we utilized McCleary's

A prominent feature of this sectioned Dart 302 Iron Eagle block is its massive cylinder-wall thickness. Even with a 4.25-inch bore, the walls are still 10 percent thicker than a stock bore. Also worth noting are the blind head-bolt holes. That means no more rusty head bolts on teardown.

Here is our Dart Iron Eagle after having been bored and honed to 4.165-inch bore diameter. You can get a bare block, but a ready-to-assemble enameled block as seen here from T&L sells for $2,700.

Part of the credit for internal balance capability must go to the less-than-average weight of the Ross pistons and Scat H-beam race rods.

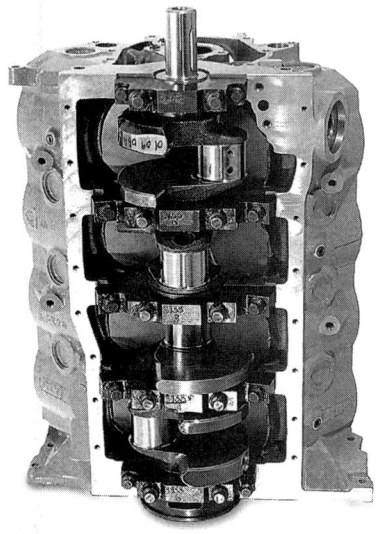

Note the massive angle bolt, four-bolt mains on the Dart Iron Eagle block and the aero counterweights of sufficient size to internally balance the entire deal despite the 3.4-inch stroke used.

considerable experience building big-inch Fords, and on his recommendation went with a Dart 302 Iron Eagle block for this step up in cubes.

The first thing you notice as the Dart block comes out of the crate is the casting quality. Because it is intended to hold out reliably against power figures well into the four-digit region, it is, at 178 pounds after boring, heavier than the stock 302 block equipped with a steel main girdle by a margin of some 46 pounds. But the considerable extra strength of this block is not just from extra material in strategic places alone. The cast iron used is a denser, high-nickel, high-strength alloy that contributes considerably to the overall integrity.

As you can see from the nearby block section, Dart has put a lot of extra material into the bore wall thickness. The intent here is to withstand big boost and/or a ton of nitrous. With walls this thick, it would be a crime not to make use of them by making a big-bore motor.

The cylinder walls are typically 0.225 inch thick. That's about 10 percent thicker than a Mexican block with a stock bore size. Add to this the stronger material and a Dart block at 1/4-inch oversize is probably still better than 20 percent stronger in the bore department than even Ford's best stock offering.

Although a 0.250-inch block looks feasible, we elected to go no more than 0.165 inch over. With the intended 3.4-inch stroke Scat forged crank, this would produce 370 inches. For the record, at Dart's

biggest recommended bore, this would have delivered 377 inches and, had we gone out on our own and bored to 4.250 inches, the cubes would have amounted to 385.

So much for big bores to make inches, but let's not forget that this is just half the equation. The other half is the Scat stroker crank and its associated longer rods. For the power level we were shooting for (550-plus horsepower), it was deemed that we were on the cusp as far as the choice between Scat's cast steel crank and the basic and cost-effective forging. At the end of the day, the forging won out by just a small margin. Even though it was Scat's least expensive crank, it still featured many of the performance assets of the more expensive NEXTEL Cup–style cranks. This included aero counterbalance weights and hollow big-end journals. Any time the stroke goes up, the amount of counterweight mass necessary to internally balance the rotating assembly goes up. In spite of this, the way Scat had done the counterweights there was enough counterbalance mass to allow for the assembly to be internally balanced.

Pistons for our big-bore engine were from Ross. These semi-custom pistons are not quite off the shelf but, as Ross claims as part of its service, the order was turned around in three weeks. These flat-top pistons feature 1/16-inch, 1/16-inch, and 3/16-inch rings. Because of the stroke and the need to cram in as long a rod as possible, the pin intrudes into the oil-ring groove necessitating a support rail–style ring and groove design. As for the rods, these were Scat's light race H-beam items. Their relatively low weight plus the light weight of the Ross piston contributed toward being able to balance the rotating assembly internally.

CAM AND VALVETRAIN

The cam profiles McCleary used to operate our 370's valvetrain were from Comp Cams. The custom T&L–spec roller profile had reasonably high acceleration rates and lift values, but required only a moderate spring, thus producing a valve

If, as in our case, the cam was well matched to the rest of the engines spec, timing it in correctly becomes more important. In this case, the intake centerline was set at 104 degrees.

The crank damper used was from Professional Products, and was conveniently marked with several degree scales. This allowed the use of any one of three different types of TDC pointers.

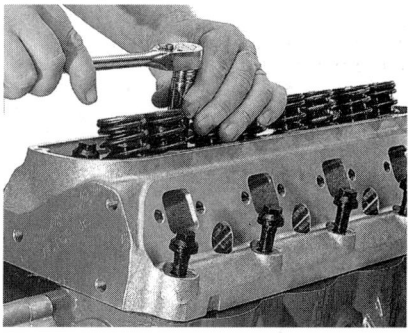

After locating the Fel-Pro head gaskets, the T&L ported Edelbrock Victor Jr. heads were bolted down using ARP's more friendly priced six-point bolts.

Comp Cams multi-keyway sprocket allowed the cam timing to be altered in 2-degree increments. In our case, the accuracy of the crank, cam, and gear set meant that the "zero" slot got the timing spot on.

There are several points to note from this shot. First the manifold gaskets are pre-glued to the heads so they do not move when the manifold is installed. Second is that the lifters must be installed before the heads. Third, the arrow shows the pedestal for the Jesel rockers bolted in place. Lastly, note the titanium retainers, which hold the Isky "tool room" springs in place.

Our Jesel Sportsman rockers looked more pro than sportsman, but the cost was definitely sportsman. Here, lash is being set at 0.016/0.018 for intake and exhaust, respectively.

train with good street/strip reliability. The advertised and 0.050 duration figures were 296/260 for the intake and 304/268 for the exhaust. With a 1.6:1 rocker ratio, gross lift was 0.672 inch. This was ground on a 108-degree LCA with 4 degrees advance. To get the cam timed right, a Comp Cams multi-keyway double-roller timing chain assembly was used. This allows the cam to be adjusted in 2-degree increments. Ours went in at the required 104-degree intake centerline. Had the actual adjustment from the multi-keyway sprocket given an option of 103 or 105, it would have been better to go slightly more advanced than retarded as this has less negative effect on power. With the cam timing buttoned up, the timing cover and crank damper were installed, and our attention turned to the Edelbrock Victor Jr. heads (see sidebar on page 56 for details) and rocker shaft assembly.

The head gaskets were Fel-Pro's high-performance items, and the T&L–ported Victor Jr. heads were held down by ARP's lower cost six-point shouldered bolts. For rockers, we elected to use a Jesel Sportsman shaft setup. These are well-known and trusted race rockers that Jesel produces with the frills removed. What's left is a respectable shaft-rocker setup at an equally respectable price. After the heads were torqued to 85 ft-lb, the pedestal base for the rockers was bolted into place. At this point, a pair of adjustable pushrods were installed, and one pair of the shaft-mounted 1.6 rockers was bolted down to the pedestal. Adjustment of the pushrod length was then made until the desired sweep pattern on the valve tips was achieved. With that the pushrods were removed and the length checked. This length was then ordered from Comp Cams. They were then installed, and the Jesel

HEAD GAMES

A set of Edelbrock Victor Jr. heads were chosen because of the intrinsically sound port and chamber design. CNC-ported Victor Jr.'s would have been a simpler choice, but with a 225cc port runner, McCleary considered they may have a small but nonetheless noticeable impact at low speed. This engine is after all a street/strip deal, and that means not overlooking the street part of such. What was done instead was to take a set of as-cast Victor Jr. heads and port them. Edelbrock rates the ports of the CNC version of this head at 320 (intake) and 220 (exhaust) cfm at 0.700 inch lift. What we wanted to see was whether these numbers on the intake side at least, could be achieved with a smaller port. The good news is McCleary made a pair of these heads go 321 cfm on the intake with a 217 cc port and 227 cfm on the exhaust of only 80 cc volume.

But peak numbers are not the real issue unless the valve is going to be lifted into that range. Our cam and valvetrain combination would, after lash, lift the valves to a little over 0.650 inch. If the port goes on flowing significantly bigger cfm numbers after peak lift, it is a sure sign the port is too big.

What really presents a truer picture of the situation is to look and see how fat the flow curves are rather than how tall. Take a look at our flow graph. This is what T&L's Superflow measured, and that is after we used the same calibration procedure as used for UNCC's flow bench. At 0.250 inch, the intake was close to 165 cfm, which is better than 90 percent of the heads out there with 2.1-inch valves and a conventional valve job. Notice how the flow tipped over just before the full amount of used valve lift was reached. This indicates that as far as valve lift is concerned, the port is almost perfectly matched to the lift used. This leads to minimal redundant port volume, which in turn means nearer optimal port velocity for the job at hand.

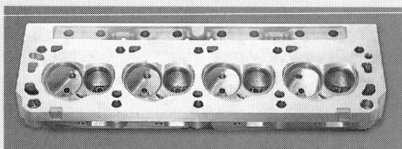

This is what the Victor Jr. looked like after getting T&L's pro porting treatment. The ports featured a brushed-like finish, not the high polish so often put on heads to impress an uninformed buyer.

After calibrating T&L's computer-supported Superflow bench, our project engine's heads were flow tested. The 321/227 cfm for intake and exhaust were actually up on ports with a significantly bigger volume.

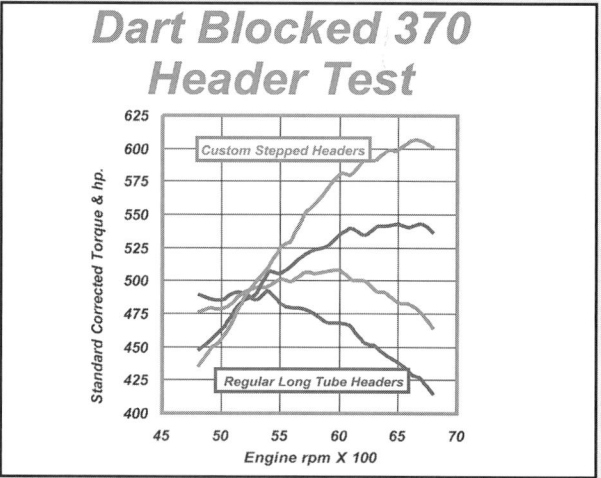

Flow bench results indicate that the T&L–ported Edelbrock Victor Jr. heads had fat flow curves. They also had excellent top-end flow while still having ports of less volume than usual. This was the sort of combination that could be expected to give a wide and torquey powerband while still delivering good top-end power.

rocker shaft assemblies positioned and bolted down. The rockers were then lashed at 0.016 inch/0.018 inch for the intake and exhaust, respectively, and with that the valvetrain was about complete.

PAN AND PUMP

It was about this time that the Moroso oil pan came in, so attention was returned to the bottom end. The oil pump used was a Melling high-volume design driven by an upgraded Melling driveshaft. The decision-making process for the Moroso pan was based on the need to have something that was good for both the dragstrip and road course. The fact the car would be lower at the front than would be the case for a drag race–only deal meant the pan would have to be shallower than normal. This makes the whole process of keeping the oil out of the way of a crank zipping around at 7,500 rpm that much more critical. It meant more attention had to be paid to keeping the pump pickup immersed in oil. To this end, a triangulated trap-door system allowed oil to flow easily into the pump pickup area, but resists it leaving by any route other than out through the pump.

INDUCTION

We chose two intakes for our 370—the Edelbrock Victor Jr. and a larger Super Victor. It was decided to test with both of these as the engine was at the crossover point where the bigger Super Victor might pay off. These were ported by T&L's ace porter, McCleary. If you've ever tried it, you know the physical job of porting an intake manifold is a lot more difficult than porting heads. The porting and matching job done on these manifolds was nothing short of top-notch. It was more than just a cosmetic clean-up job, as attention had been paid to progressively profiling port areas

We checked out T&L's port-matching job and it proved to be right on the money both width-wise and top to bottom. The manifold was secured with stainless washers and 12-point bolts.

Always pay special attention to minimize potential windage losses by selecting an appropriate performance pan.

Seen here is the AED-modified 750 Holley carb, which after modifications flowed nearer 825 cfm. Check out the plug wire harness. Examples like this can be had from Moroso or Mr. Gasket.

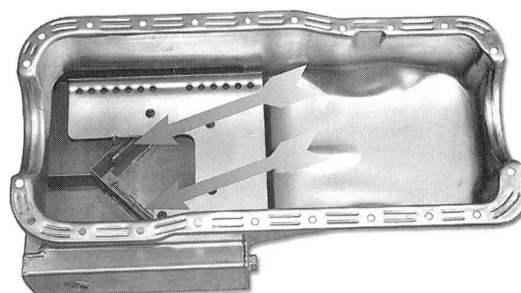

Here is the Moroso pan used on our project, which could find itself on either a dragstrip or road course. The arrows indicate the two one-way trap doors in the oil pump well of the pan.

The workmanship on this T&L ported intake had to be seen in the flesh to be appreciated. The porting was not just cosmetic, but sought to achieve the best port tapers for the job.

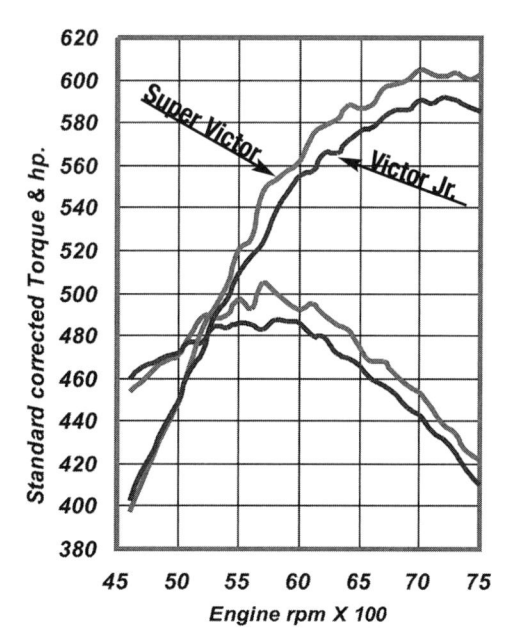

Dart 370 Intake Manifold Test. Victor Jr. vs. Super Victor.

Super Victor

Victor Jr.

Standard corrected Torque & hp.

620 600 580 560 540 520 500 480 460 440 420 400 380

45 50 55 60 65 70 75

Engine rpm X 100

Any carbureted small-block Ford hovering around the 580hp mark is a candidate for testing a Victor Jr. against its bigger sibling, the Super Victor. Under 5,000 rpm, the Victor Jr. was best with up to 10-12 lb-ft more torque. When rpm passed 5,000, the Super Victor came on strong, bumping both peak torque and horsepower by 15 numbers.

Crane's all-electronic distributor has a combined total of 36 rpm and vacuum advance curves built in. These are selected by turning the two indicated screwdriver slots to a number corresponding to the curve required.

from the plenum out. The goal was to make the port area get smaller from plenum to head at a rate equal to about 1 1/2 degrees. In addition to this, attention has to be paid to the way air enters the runners.

For its cubes and rpm band, a carb of about 825–850 cfm was deemed necessary. The easiest way to fit this bill would have been to use an off-the-shelf 850 Holley. But McCleary wanted a carb that would produce the best low-rpm results for street driving, yet flow well enough to make good top-end output. His choice, based on dyno results, was an AED-modified 750 Holley.

When Holley built the 750, it came up with about the most versatile platform within the configuration limitations of a four-barrel carb for small-block V-8s. AED builds on that with its street/strip 750. Without impairing the low-speed characteristics, AED reworked the carb and picked up about 65 cfm. This is done without increasing the size of the main venturi. The result is a carb that has the low-speed characteristics of a 750 but the power capability of one that flows about 825 cfm.

WE HAVE IGNITION

For ignition, we went with something a little different. Here a Crane distributor, with built-in electronically generated advance curves, was used. An internally housed chip has, programmed into it, nine different curves that advance with respect to rpm. That is what we would have called mechanical advance on a conventional distributor—except there is no mechanical device present to carry out the routine.

For our terminology to be correct with a Crane distributor, we will have to refer to it as rpm advance. To allow for the characteristics of various cams from mild to wild, one of three different built-in vacuum curves can be selected. Accessing either rpm or vacuum related curves is as simple as turning a selector to the appropriate number (how much easier can it get?).

Dyno Time

Our Victor Jr.–equipped 370 went from engine stand to purring on the dyno with about 30 minutes work, and, equipped with regular long-tube Mustang headers, was ready for the first round

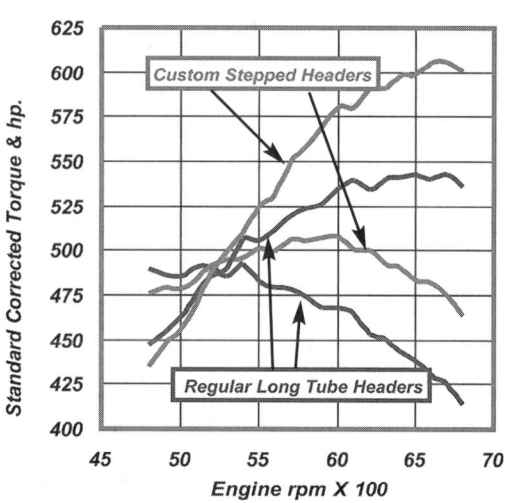

DART BLOCKED 370 HEADER TEST

Custom Stepped Headers

Regular Long Tube Headers

Standard Corrected Torque & hp.

Engine rpm X 100

Trick Exhaust—What's It Worth?

One aspect about building competitive NASCAR engines is that with rules restricting output for various classes and track to levels between 430 hp and over 800, is that you get to have a good handle on what a near optimal exhaust should be for any given power level. This was just the case here. McCleary reckoned a typical Mustang long-tube header was more suited to engines in the 350-450hp range. This prompted him to use a system having dimensions more appropriate to a 600hp engine. The 1 3/4-inch stepped design paid off. as the graph here shows. Gains started at 5,200 rpm, with peak torque and power climbing by no less than 15 lb-ft and 48 hp.

of testing. Since T&L has done a number of similar engines in the past, it didn't take long to set up the motor once it had been given a full two-hour break-in. The distributor advance curve was already a known factor, so it was set, as was the vacuum. This left some minor jetting work on the AED-modified Holley carb and the job was done.

The first series of tests showed the headers were not really up to the job of dealing with an engine of this power level. If you really want to make the most of an engine like this, you can forget those 15/8-inch long-tube headers. McCleary went to a 1 3/4-inch stepped header with more appropriate lengths and collector design (see sidebar for results).

The dyno chart shows the results with the Victor Jr. after the new headers were installed. The first point to note is that this was not a top-end-only engine. It would run smooth and pull like a stock

302 down in the 2,000-rpm range, but by the time it hit the 2,700-rpm mark, it started to show all the signs of being a big-cube torque monster. For a pump-gas-burning motor of nominally 10.5:1 compression ratio, the torque output of 490 lb-ft is stout. Any time the pound-feet per cube goes over the 1.3 mark, you can reckon the engine builder has done his homework.

When constraints are put on the compression ratio, it becomes more difficult to get big torque numbers. A good torque number for a 10.5:1 Two-Valve pushrod engine is 1.25 per cube. With 370 cubes, 462 lb-ft would have been an acceptably good number, so 490 was more than such. As for the horsepower, it checked in at 593 at 7,200 rpm.

At this point, the Victor Jr. intake was replaced by the Super Victor and, as we were to find, the deep-breathing, T&L-ported Victor Jr. heads loved it. The bigger manifold lost 10–12 lb-ft below 5,000 rpm, but both peak torque and horsepower above this point took a worthwhile jump. What our pump-gas burner finally delivered was a whopping 505 lb-ft of torque along with a best pull of 606 hp.

CONCLUSION

Because everything was done in a sound engineering fashion, the extra cubes afforded by the Dart block/Scat crank/Ross piston combination paid off big time. Remember, we started with the 302 configuration at the beginning of our big-cubes buildup. At this point, adding much more stroke to an 8.2 deck height block would mean dealing with a very short rod/stroke ratio combination. Although we are pushing the boundaries of diminishing returns, it would seem like we are not quite at the limit of positive returns for added stroke if torque and power output below 6,500 rpm is the goal.

If we wanted to get more cubes at this point we could still go up a tick on both stroke and bore. That, for our hydraulic cammed, street driver is where we will go for the next big-inch 8.2-inch block build. This will stretch our 370 to 382 inches.

So what did all this cost? Certainly the Dart block was the biggest investment factor here.

But having said that, a block of this obvious quality, strength, and bore-diameter capability was still a great buy. From T&L, as an upgrade over a stock block, it was $2,200. As for overall cost, a copy of this engine from T&L would set you back $9,800.

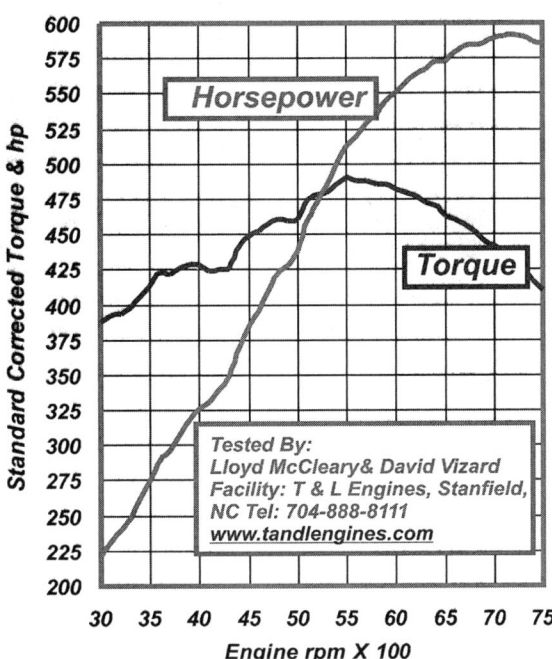

Dart Blocked – SCAT Cranked 370-Inch Stroker Dyno Test

Tested By:
Lloyd McCleary & David Vizard
Facility: T & L Engines, Stanfield, NC Tel: 704-888-8111
www.tandlengines.com

With the Victor Jr. intake, our 370-inch stroker cranked out a respectable 593 hp and 490 lb-ft of torque. It did this and still produced more low-speed torque than a stock 302, so driveability will be more than acceptable. All this was done on 93-octane pump fuel.

Chapter 9

RESURRECTING A BLOWN 5.4L

After Ventilating The Block on our Lightning Project Truck, JDM Engineering Rebuilds It Better Than Ever

Text and Photos by Evan J. Smith

It was a brisk evening back in April 2003 when I staged the Fridge at Old Bridge Township Raceway Park in New Jersey. A few of my *MM&FF* staffmates were on hand making laps in their own cars, and I decided to join in on the fun. Up to that point, our '99 Lightning project had shot to a best of 11.78 at 112 mph, but that was in race trim—meaning I had the intake iced, slicks and skinnies bolted up, and all unnecessary weight removed. On this night, I was rolling on Nitto 555 Extreme Drag tires, I had the weight in, and I was going to run with the engine at normal operating temperature.

The cold evening air made for a cold track, and I wanted to get away clean—so I eased into the throttle with a gentle touch. Despite the gross tonnage, both truck and I accelerated quickly and by the 60-foot marker, I had the gas planted to the mat. Second gear hit with a bang, and the tires barked as they fought for traction. The Eaton blower was singing that happy horsepower song, and the tach shot up to redline.

By half-track, the truck snapped into third gear and in that instant, the engine shuddered badly. I jerked my foot off the throttle, jammed the shifter into neutral, and clicked off the ignition as quick as I could. Too late. The rear view mirror filled with death smoke looming behind—trailing me like a demon.

Even with no power, I had enough speed to clear the traps and make the turn off. Sadly, after coming to a stop, I could smell burning oil and see running from the bottom of the

Our '99 Lightning blazed a trail (of oil) through the pits at Old Bridge Township Raceway Park during a drag session. Jim D'Amore assessed the damage and said, "I can fix it."

Ha, ha, go ahead and laugh. As Arnold said, "I'll be back!" Here K.J. Jones chuckles while I show off a wrist pin and a chunk of aluminum.

Back at the shop, the guys at JDM Engineering yanked the hurt pup and found that I did indeed chuck a rod (or two, or three) out the side(s) of the block.

That's one serious window.

We didn't have to remove the piston from this bore because it was gone when we took the heads off. I've never seen a block with cracks the entire length of the bore.

No comment.

truck. That's going to leave a mark. I was really bummed; then a track worker pulled up and said, "I think this came from your truck." He then handed over a wrist pin—as if I could replace it and get back to the lanes. "Thanks," I replied.

My attentive workmates soon realized I hadn't returned to my space and came to my rescue, but there was not much they could do. We towed the F-150 to the pits, and I called Lightning guru Jim D'Amore. Within the hour, our truck was on D'Amore's open trailer en route to his popular vacation destination, JDM Engineering, located in nearby Freehold, New Jersey.

It didn't take a rocket scientist to figure out that the damage was terminal. With a flashlight I was able to see the crankshaft through the new window in the block. The damage had severely bent one of the header tubes.

After returning to the shop, we got to work and pulled the 5.4 supercharged engine from the bay. I was curious to see how bad the engine was damaged and even more curious about why the failure occurred. We'd been running the same tune-up for over a year, and I always run 93- or 94-octane fuel. And contrary to popular belief, I don't beat the truck hard. Sure, I run it severely on the track, but I baby it on the street.

Anyway, the guys at JDM had the 5.4 out and that made it easy to see the monster hole in the block (see photos) on the driver side and a small hole on the opposite side. In addition, there was one rod missing in action; another had the beam and small end missing, and one was bent like a pretzel. Three pistons were smoked, and one of the cylinders had two giant cracks extending from top to bottom. This thing went out Top Fuel–style for sure!

Okay, so you know the story about the death; now let's get to the resurrection. There weren't many parts left to work with so D'Amore started with a '03-model engine block and crankshaft. The block was bored .020-inch over, and he installed Probe forged aluminum pistons that are full-floating and swing from Manley/JDM H-beam rods. The Manley rods can handle 700 hp and up to 8,000 rpm. They are forged from 4340, have ARP bolts, and are bushed. The pistons are made from 2618-T6 aluminum and have an 18.1cc dish. A stock crank was balanced to the rods and pistons and installed in the block, along with a replacement oil pump and a stock pan.

Our new engine would benefit from Probe forged aluminum pistons.

Since no aftermarket heads are available, we took a stock set and had them hopped up by M2 Race Systems in Farmingdale, New Jersey. M2 CNC-ported the aluminum casts and installed Ferrea valves, Crower springs, and Crower billet camshafts ground to D'Amore's specs. The cams are rifle-drilled and feature .581 lift and a little extra duration. (If you want to get more specific, ask D'Amore. He wouldn't tell us.)

The heads were attached using ARP bolts, and then the long-block was topped with a Magnum Powers supercharger. The MP unit looks similar to a stock Eaton, but there are some subtle changes that will give us some extra power. For starters, the case was designed so that the rotor pack is moved forward about 3/4-inch. This opens up the inlet

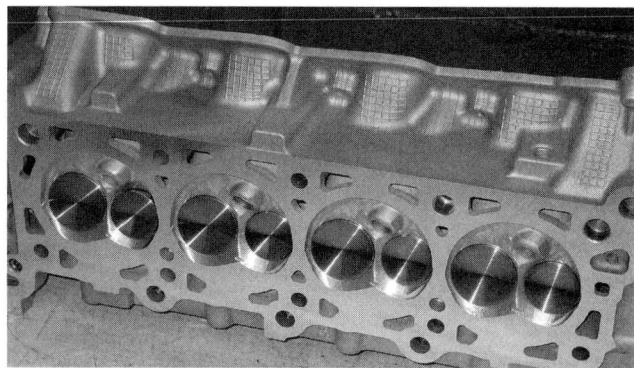

The cylinder heads are stock casts that have been CNC-ported by M2 Race Systems. According to Ron Mielbrecht of M2, the heads are ported on a five-axis machine and are flowed prior to delivery.

and increases flow. Looking underneath you'll note that the MP blower has a much larger outlet and the edges of the V-shaped opening are smoothed to further enhance airflow. Lastly, the four smaller holes found on the stock blower are removed. D'Amore stated that the smaller holes on the Eaton are used to quiet the unit down and removing them helps performance.

D'Amore bolted down the blower and installed one of his JDM phenolic upper intake spacers between the blower and the upper inlet plenum. Finishing off the package is a big single-blade Accufab throttle body and 55-pound injectors that replace the stock 42-pound squirters. He then installed a new set of JDM 1 5/8-inch headers, but switched to a JDM H-pipe and mufflers.

In true *MM&FF* fashion, we went directly from the lift to the dyno for tuning and wide-open power pulls. D'Amore broke the engine in by getting it up to operating temperature, and then he got to the task of tuning. D'Amore has become an authority on tuning Lightning engines, and what he saw on the Dynojet scared him nearly to death. What could it be, you ask? He saw an incredibly lean air/fuel ratio in the 14.7:1 range, and that really got his attention. A check of the chip revealed that all was well in computerland. The engine should have been getting the proper fuel, but it wasn't.

Eventually he diagnosed the problem as a faulty fuel pump relay that prevented the fuel pump from supplying full flow. He was also confident the bad relay caused the engine to blow in the first place. Good thing he caught it or we would have popped another mill right there on the dyno.

Once the relay was replaced, the engine came to life. D'Amore upgraded our 4-pound pulley kit with one of his 6 pounders and The Fridge developed mad power and 16 psi of boost. Horsepower topped out at 542.5 at 5,200 rpm and

According to D'Amore, the ring package has been moved down, but the 18cc dish has been retained. The Probe pistons are very quiet and will withstand the power level we plan to achieve.

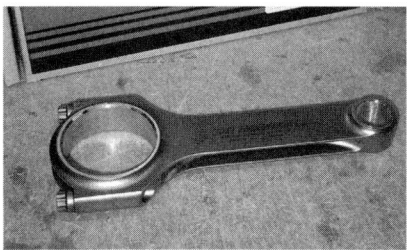

They will be replaced by Manley/JDM H-beam rods designed to handle up to 700 hp.

ARP main studs hold the crank in place.

The stock powdered metal rods exited stage left.

JDM Engineering used an '03 block that's been bored .020-inch over. The rotating assembly was balanced and carefully assembled using ARP hardware, Probe pistons, and Manley/JDM rods.

The 5.4 Triton has a long 4.1-inch stroke, and that's one reason these engines make Earth-moving torque.

Intake valves measure 45.5 mm (1 mm larger than stock), and peak flow through the intake port is 232 cfm at .800-inch lift. We'll be using a .581-inch lift cam, so we're more concerned with the flow numbers up to about .600-inch lift. Flow at .500 is 210 and flow at .600 is 221 cfm.

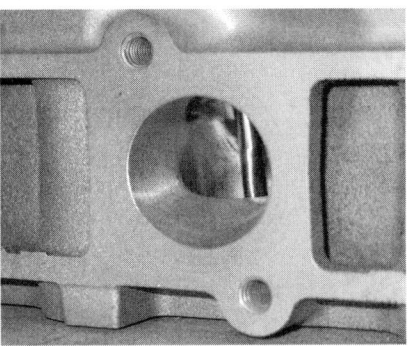

The exhaust valve is 37 mm (1 mm larger than stock), and the exhaust port flows 162 cfm at .500 and 171 at .600 cfm.

JDM chose a billet Crower cam that's been rifle-drilled and features .581-inch lift.

This is a stock Eaton 112 blower.

Our new blower is from Magnum Powers; it is not an international man of mystery.

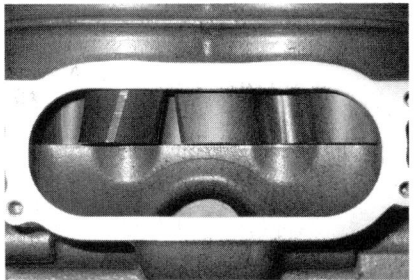

You can see how the rotors in the Eaton blower sit halfway in the opening.

It looks close to stock, but it's not. The Crower cams give the Lightning a nice rumble at idle, but drivability has not been compromised.

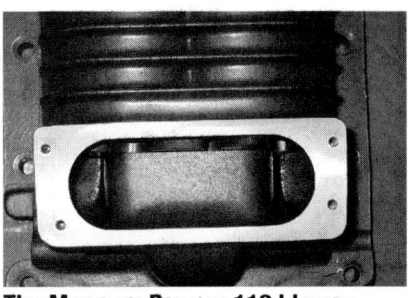

The Magnum Powers 112 blower uses stock rotors that have been moved forward in the case. This opens the entry to provide more flow. This early Magnum blower does not have a boost bypass, but the production models do.

Underneath there's quite a difference between the Eaton and the Magnum blowers. The V-shaped outlet is much larger on the Magnum, and you can see how the trailing edge has been smoothed. The four smaller holes have been filled in and the lip is rolled.

peak torque was 635 lb-ft at 3,700 rpm. Our blown 5.4 made over 500 hp from 4,350 rpm to redline, which was set at 5,800 and torque was over 550 lb-ft from 3,100 to 5,200 rpm.

The numbers speak volumes about the performance potential and, from our initial road testing, we can tell you it's one strong mother. Next we'll hit the track with slicks and head for the low 11s. D'Amore stated that he can dump in some race gas and turn up the boost, and we're ready for it. We don't like to make predictions, but I'd be really happy if we ended this ride with a 10-second timeslip.

MORE TESTING AND TUNING

"When we start tuning a new engine or a new combination, we fire it up and get the engine and all the fluids up to full operating temperature," stated D'Amore of JDM. "I like to see about 192 degrees of engine coolant temperature, 100 degrees on the inlet air temp and 125 degrees on the intercooler coolant on Lightning trucks. If everything sounds and feels good, I'll make a half-pull, from about 2,500 to 4,000 rpm, then I look at the data. We always start with a base tune-up in the chip. On Lightning trucks the chip has about 2–3 degrees of timing over stock.

"The stock timing is 11 degrees and the target air/fuel we shoot for is 11.8:1 at WOT. During

JDM dyno-tested the truck after 1,000 street miles. With the same tune-up as before, the 5.4 achieved 16-psi of boost and the result was 559 hp. BOTTOM: Surprisingly, peak torque dropped from 635 to 628, but the average from 3,500–5,500 rpm climbed by 6 lb-ft.

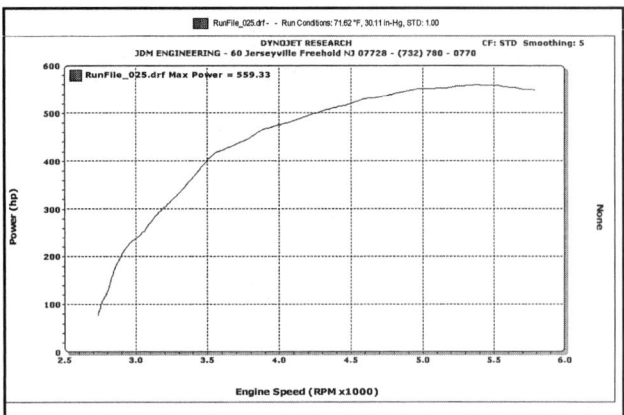

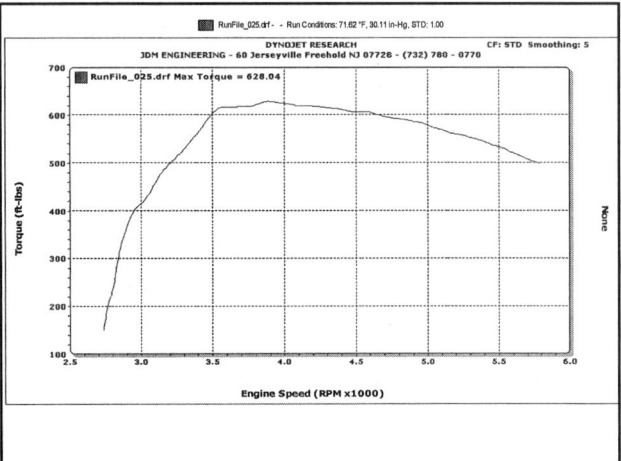

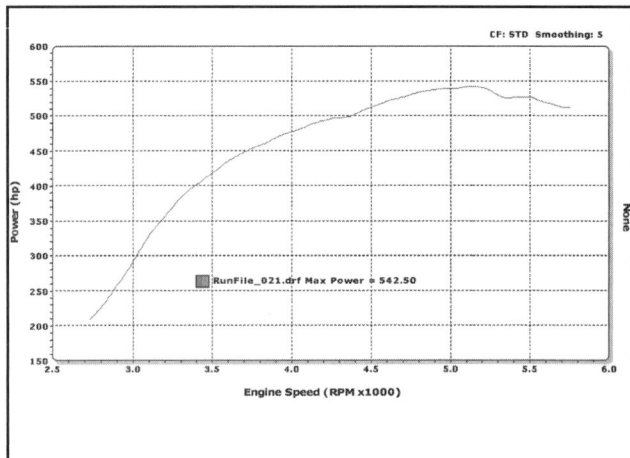

Peak horsepower at the tires was 542.50, exactly 101 more than we had before. We expect to run low 11s, but we're really hoping for a 10.99.

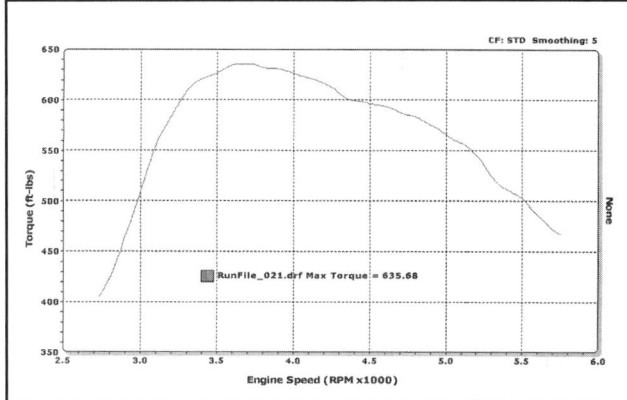

Got torque? With 16 psi of boost, Fridge cranked out a whopping 635.68 lb-ft of torque at 3,700 rpm. Amazingly, the Triton puts out over 500 lb-ft from 3,000 to 5,500 rpm!

testing I monitor fuel pressure, timing and the a/f. I don't mess with the fuel curve too much at all. It's pretty good from Ford unless you go past five volts on the mass air meter. If it hits five volts the computer goes to a default table or map and that is too lean for most modified engines. The computer doesn't know there is 100 or 200 extra horsepower and it can't adapt in the default tables," added D'Amore.

As we mentioned in earlier, D'Amore discovered that the relay controlling the fuel pumps was bad and that caused a severe lean condition and the subsequent engine failure. Once the problem was fixed the truck was run hard, first making 495 hp with a 6-pound pulley and a base tune-up. By the end of the session we had the 8-pound pulley in place and The Fridge pumped out 542 hp and 635 lb-ft torque.

With the newfound power the truck was a complete animal, in fact any attempt to put the power down resulted in a dramatic tire spin—even from 40 miles per hour. Drivability was of OE quality, but with an aggressive rumpity rump at idle. Oh, and lest I forget, there is an authoritative whine from the Magnum Powers blower when you crack the throttle open. I drove the truck for a few

days and you couldn't get me out of the seat.

AT THE TRACK

With all this extra power on tap we had one thing in mind and that was to get to the dragstrip. We would have made a beeline from the shop to the track, but I had a prior engagement that required me to be out of town so the test had to wait. I utilized the down time by dropping the truck off at JDM Engineering so they could change the oil and go over the truck before I raced it.

D'Amore took some extra time and dyno'd the truck once again, figuring it would make more power now that it had 1,000 miles on the new engine—and it did. With no additional tuning the 5.4 spit out 559 hp and 628 lb-ft of torque. Horsepower was up by 16 numbers at the peak and an average of 9 hp from 3,500–5,500 rpm. The torque peak was down 7 lb-ft, but the average in

Traction came by way of a prepped dragstrip and our Goodyear 30x9-inch radial slicks. Believe it or not, we ran with 17-psi of air pressure in the tires. The radials are normally worth a tenth over similarly sized bias-ply slicks.

Notice that the slicks barely wrinkle up on launch. The radial design reduces rolling resistance, thus reducing friction and parasitic drag.

the 3,500–5,500 range was up by 6 lb-ft.

Actually, I didn't care how much power it made; I was ready to run. And upon my return I picked up the truck, along with a set of slicks and skinnies, and made my way to the track. We set the truck up for racing by removing the passenger seat and bolting up the tires. For improved traction we opted to leave the tailgate and the 80-pound spare tire in place.

If there is one neat thing about the Lightning trucks, it is they are easy to drive fast. To gain maximum straight-line performance we enhanced our truck with QA-1 rear shocks, Metco traction bars and we removed the front anti-rollbar.

When making a run I use the following procedure: First and foremost, I always run with the engine cool. So once the engine temp is down I fire the engine, turn off the overdrive and put the transmission in "low." I prepare to do the burnout by wetting the tires, then I roll out of the water puddle, just past the leading edge of the water, but not too far into the sticky stuff. I begin the burnout in first gear, then I shift to second and hold it until the tires smoke a little. When the tires begin to

smoke I drive it out, feathering the throttle to keep the revs below 5,000 rpm. I don't recommend a dry hop, I just put the transmission in drive and prestage.

Once prestaged, I press hard on the brake with my left foot, and bring the engine rpm up to 2,000 and ease off the brake pedal to let the truck gently creep into the stage beam. My attention then goes to the Tree and on the last yellow I release the brake and jam the throttle to the mat. If the track is working well I plant the gas as quick as I can. If the track is loose I squeeze the throttle quickly, but in a smooth manner. Lastly, I hold on and enjoy the ride.

My first pass with the new engine resulted in an 11.25 at 118.90 mph. Not quite in the 10s, but not a bad start either. The weather on this particular day was not racer friendly, as there was lots of humidity (90 percent), the barometer was 29.40 and the ambient temperature was 85 degrees.

On the first run I underestimated the traction and didn't launch hard enough, so I went up for another run and dropped the hammer. Planting the throttle hard and fast, The Fridge leaped forward, the 60-foot time improved by .03 from 1.60 to 1.57 seconds and the run was over in 11.21 seconds. My trap speed was virtually identical to the run before at 118.87 mph. A third baseline run was made and the Lightning cruised to an 11.20, but this time the trap speed was up to 119.22 mph. The gain in mph was attributed to the improved weather conditions and the fact the head wind had died down considerably.

With three consistent runs under our belt we decided to get after it by icing the upper intake. D'Amore was ready to throw a hotter tune-up at it, but I wanted to make a few more runs with the street tune. "I set The Fridge up with 13 degrees of timing, but I can go up if the truck is only going to

Jim D'Amore takes a look at his creation. It's pretty amazing to think that you can have a 4,400-pound daily driver that runs low 11s or high 10s.

The Fridge is powered by a JDM-built 5.4 with a Magnum Powers blower, M2 ported heads, Crower cams, JDM exhaust and tune (with a Superchips chip), Probe pistons, Manley rods and Accufab throttle body. Other mods include twin 255-lph fuel pumps and MSD 50-pound injectors. The Magnum Powers supercharger is a very cost-effective upgrade from the stock Eaton; some blowers have the potential to make more power, but the entry price will be significantly higher.

During the burnout it's important to keep the rpm in control, because you don't want to over rev the engine and bang the rev limiter. I drive it out under power and try to get out of it just before the tires grab.

be used on the strip. The extra timing advance will really increase the power, but I don't recommend driving it on the street like that. I don't like to push it much past about 470–480 rwhp on 93 octane," added D'Amore.

"When we went for max power with The Fridge we had a 50/50 mix of 93- and 100-octane race gas, but the chip is still okay to be used with straight 93 with a 4- or 6-pound pulley. We will run the 8-pound pulley in the controlled environment at the track, but not on the street. Basically, I like to talk to each customer, because everyone has different driving habits and needs," explained D'Amore.

After a 40-minute cooldown we dropped the hood and dropped the hammer. I heated the slicks, staged shallow and torqued the engine up to 2,200. I let it rip and the F-150 squatted and dug in hard. The blower made that awesome sound and at 5,600 rpm the transmission snapped into second and then third. I knew by the 1.53 60-foot time (which was posted on the scoreboard) I was on a hot lap. Then, suddenly, as I approached the traps the transmission dropped into overdrive and the engine pulled down hard. "Crap, I forgot to cancel overdrive," I muttered to myself.

The torque kept the Lightning hauling and just a second later I crossed the stripe. Not all was lost as the clock flashed my 11.15/116.50. It was our best run by a ton, but the overdrive killed the speed and a little elapsed time. While we didn't achieve maximum performance, I feel we only lost about .02–.03 at the most.

Naturally, we had to give the rig one more try so on went the ice and 40 minutes later I was back at the line. I employed the same burnout and launch technique as before, albeit with the OD switched off this time. With mounds of torque the truck left the line with aggression and the 60-foot time was 1.52, but the e.t. slowed to an 11.17. Of course, the slower e.t. was disappointing, but that's the way it goes.

All in all, we're quite pleased. JDM prepared the

SOURCES

Crower
3333 Main Street
Chula Vista, CA 91911
619/422-1191
www.crower.com

JDM Engineering
60 Jerseyville Avenue
Freehold, NJ 07728
732/780-0770
www.teamjdm.com

M2 Race Systems
5140 West Hurley Pond Road
Farmingdale, NJ 07727
732/751-1902
www.M2race.com

Magnum Powers
51850 Northwest Outback Lane
Forest Grove, OR 97116
503/357-5444
www.magnumpowers.com

Manley Performance Products Inc.
1960 Swarthmore Avenue
Lakewood, NJ 08701
732/905-3366
www.manleyperformance.com

Probe Industries
2555 West 237th Street
Torrance, CA 90505
310/784-2970
www.probeindustries.com

truck just as we asked. Now we have enough power to run hard, but with enough manners to drive it everyday. As we stated earlier, the tune-up in the Superchips chip is the one we use on the street, but we can lean on the engine by swapping pulleys or a more-aggressive chip. For daily driving, D'Amore recommends the 4- or 6-psi pulley, but for track use he recommends the 8-psi cog. And JDM's interchangeable system allows a pulley swap in a matter of minutes, making it the perfect upgrade for any street/strip truck.

Chapter 10

BOLT-UP
5.0 POWER

Text and Photos by David Vizard

Here is our UNC Charlotte—assembled 5.0 in finished form. Even though it was a simple spec, results were close to spectacular.

Our bolt-together 5.0 power project is being assembled by a group of students from the University of North Carolina at Charlotte's motorsports program, who put together a 5.0 using a set of ported aluminum heads with 1.94/1.60 valves. The idea was to use these heads and, with the help of the good guys at D.S.S., build a long-rod street motor using the company's billet lightweight 5.4-inch rods. With a 270/280 Comp Xtreme flat-tappet hydraulic cam and an 11:1 compression ratio, this test engine made 370 lb-ft and about 415 hp. There was, however, a lot of engine-component prep, such as porting the heads, which could be done at UNCC but not necessarily by most guys in the garage at home.

LONG-BLOCK LOWDOWN

The long-block for our project was essentially the same D.S.S. block as was used with our initial build. All that was done here was to strip and clean the entire bottom-end parts. To save money, the D.S.S. main girdle and windage tray were left out of the build this time around. This may have cost us a little top-end power, but it was accepted as a casualty of cost cutting. Except for the deleted windage tray, everything else went together as per the original build, including the Calico-coated D.S.S. pistons, and Total Seal rings along with Calico-coated rod and main bearings.

Upon inspection, the Total Seal rings from the previous build had worn no more than whatever it took to break them in. This being the case, they went straight back in the bores from which they originally came. On the front of the crank, which was the earlier and sturdier 28 oz/inch item, a D.S.S. Power Bond crank damper was installed.

THE VALVETRAIN

From the previous episode of our D.S.S. long-rod build, it was shown that for a shorter-period cam, a flat-tappet design can actually deliver more area under the curve. The point where a roller's area superiority takes over is about 275 degrees of seat timing. In this bolt-it-together phase of our

After hot tanking, bore-glaze busting, and painting, the D.S.S. block from our previous project looked just as it did new from D.S.S.

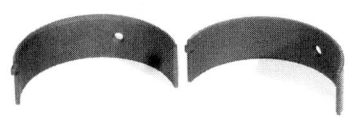

The wear resistance of Calico-coated bearings during dyno and race conditions has led to our regular usage of the parts.

The D.S.S.–prepped crank used was the earlier heavier 28-oz/inch design as, in use, it has lower main bearing loads.

By keeping weight as a consideration, the D.S.S. piston comes in at significantly less than most other regular off-the-shelf pistons. Our Calico-coated pistons were from the last build and cleaned up as new.

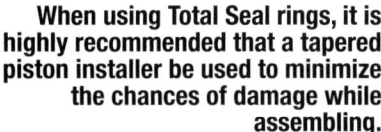

When using Total Seal rings, it is highly recommended that a tapered piston installer be used to minimize the chances of damage while assembling.

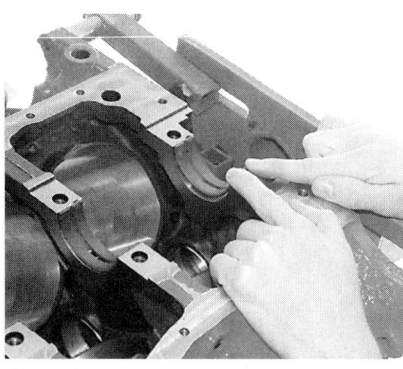

When installing main or rod bearings, push in on the side to prevent scraping the bearing OD and creating a possible unseated condition.

D.S.S.' billet 5.4 rod is about the same weight as the stock 5.0 rod and is good for 8,000-plus rpm. The extra length cuts cylinder-wall loading and bore friction.

The pan used was a basic D.S.S. street/strip unit with a single baffle. This, plus the extra capacity from the kickouts on either side, make it good for straight-line or cornering competition.

build, we wanted to be a little more traditional in terms of a 5.0 cam selection. This meant using a hydraulic roller.

Allowing the fact that this had to be a totally streetable motor meant we had better not go overboard on duration. With this in mind, a search was made through Comp Cams' hydraulic roller profiles. Throwing everything into the equation, including the flow characteristics of the heads to be used and the need for respectably high lift, resulted in the selection of the intake and exhaust profiles. The intake profile was 280 degrees of off-the-seat duration and 284 on the exhaust. Both were 224 degrees at 0.050 inch. Another consideration was that the intent was to run with the springs the heads came with. This meant the agenda included investigating profiles that were easier on the valvetrain than, say, Comp's Xtreme series. With the 1.6:1-ratio Crane rockers, which actually measure out closer to 1.65, the measured lift came to 0.575 on the intake and 0.545 on the exhaust.

With the profiles selected, we went ahead and ordered our cam, which was to be ground on a 110-degree LCA with 4 degrees advance, together with lifters and a new spider and dog bones.

TIMING THE CAM

Having diligently selected a cam that is essentially a known performer, it makes sense to time it so that it actually delivers the intended valve event timing. Another factor you should know is that a cam with too much retard has more negative impact on the power curve than one with too much advance. We had to shoot for 4 degrees advance, but 5 or 6 would still be OK, whereas 3 would not. To achieve the desired timing, we used Comp's nine-keyway

Our 280/284 hydraulic roller cam was installed with a new spider and dog-bone kit from Comp Cams.

sprocket set. This gave us the ability to adjust in 2-degree increments. By using this timing set, the called-for 106-degree intake centerline timing was precisely achieved. At this point, the valvetrain for the No. 1 cylinder was assembled, as was the optimum pushrod length to produce a centrally placed sweep over the valve tip. With that knowledge, a call was put into Comp to ship us the requested length.

The next valvetrain item we need to talk about is the rockers. Crane rockers were selected because our own tests had shown they had the geometry to lift the valves a little quicker off the seat than most others out there. This is important. Remember, we are now using a roller cam, which has less acceleration in the first 7–10 degrees than a flat tappet. This is the downside of a roller, but it can be compensated for in part by making an informed choice when it comes to rockers. Crane 1.7:1 rockers were considered here as we are looking for all the lift possible, but they would have put the system dangerously close to spring-coil bind. Our second choice was the 1.6, and this is what we went with.

HEADS AND INTAKE

Obviously, before the top half of the valvetrain could be installed, the heads had to go on. We used Edelbrock Performer RPM heads, and the justification for using these is covered in detail in the sidebar on page 70. As usual, our head gaskets of choice were from Fel-Pro, as were all the other gaskets. As with our previous build, the intake chosen was an Edelbrock Performer RPM Air Gap. It delivers the bottom end Edelbrock says it will, but given enough carb capacity, much more top-end output and rpm than the company claims. Based on our dyno testing, our recommendations for carburetion are a 650 for cams up to about 275 degrees duration and a 700–750 for 280 or more.

Remember, before you take these recom-

mendations as being universal, they apply to a dual-plane intake, which has substantially different requirements than a single-plane unit. For carburetion, we started the ball rolling with our trusty 750 vacuum secondary Road Demon from Barry Grant. As you can see from the lead shot of the finished engine, the induction system was complemented by an AED throttle linkage and bracket. If you want to really spoil yourself and get a smart, functional, throttle linkage into the bargain, this is what you should use.

CYLINDER HEADS

The choice of heads for our project engine was not something that came about simply by hunting through catalogs. The UNC Charlotte flow bench was put to good use, and the results of many (but not all) of the popularly used heads were compared. One factor that played a strong role was the out–of-the-box flow-bench performance in relation to price. Here, the Edelbrock Performer RPM heads scored well. Not only did they closely replicate the flow figures of our smaller valved but ported Windsor Jr. Lites from World, but they also delivered some of the highest swirl numbers of any 20-degree Ford head. All these assets made them look good for our purposes.

A check with a few well-known engine builders who have great experience with a wide variety of heads sealed the deal, and a set was duly obtained from Edelbrock. The fact that the heads have high swirl means that better low-speed output is likely. This in turn allows a slightly larger cam to be used without the low-speed output becoming unacceptably low for the street. In practice, we found we could use a 280-degree-seat duration cam and still come up with a better low-speed output than an otherwise stock engine with a much smaller cam.

The Performer RPM heads from Edelbrock feature intake ports accurately located in the stock position so as to best facilitate port alignment.

Other than the 2.02/1.60 valve combination, part of the good flow from Edelbrock's Performer RPM heads is due to the hand-blended bowl work.

Apart from good airflow characteristics, the Edelbrock heads also featured quality hardware in the form of good studs, guideplates, valves, and stem seals. Of great importance, though, was the use of a good spring and damper combination. These delivered a seat force of 90 pounds and 240 pounds at 0.500 lift. With its 300 lb/inch rate, the over-the-nose force with our 0.575 lift valvetrain was 262 pounds.

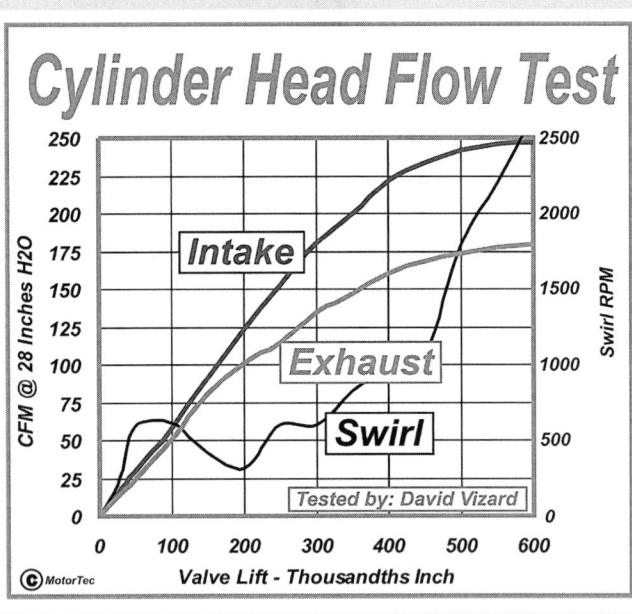

Although flow for an out-of-the-box head proved to be good, the Performers' real strong point proved to be its high swirl. Most other 5.0 heads suffer the same problem as the stock heads in that they lack sufficient swirl to maximize low-speed output. Because the swirl really takes off at higher lift, it is best to use, as we did, a valvetrain that can utilize this.

Edelbrock made the best of the stock port location by paying close attention to port form to maximize flow. These heads have enough flow to make good use of a nitrous kit.

IGNITION SYSTEM

In most cases, if we see the opportunity to make life simpler and still achieve the goals at hand, we take that route. Using a Performance Distributors, Ford-adapted, GM H.E.I. does just that. These distributors come from Performance Distributors with a custom advance curve that is just right for the job. In addition, it is capable of delivering a race-intensity spark to about 9,000 rpm in a manner as precise as we would all like. Lastly, and probably most importantly, it's a one-wire hook-up deal. Just connect the red wire to 12V on the positive side and your entire ignition system is ready to go. Now how simple can that be?

As for plug cables, we usually use Accel's bright-yellow wire wind 8mm race stuff. When our engine builder, Mike Keena-Levin, got around to the plug-to-distributor wiring, it was found we had only the Accel super-high-temperature heat-resistant, black cables in our stock of parts. Now you may well ask where the snag is here—that is, if you even thought there was one. The truth is that Accel's yellow cables are more than up to the job—and are photogenic, to boot. Our Accel high-temperature black cables are known to do a great job on the dyno. Here, the exhaust heat is far more destructive to plug boots than when the engine is installed in a car. If your budget allows, consider running these high-temperature Accel cables, especially if you have headers really close to the boots. In all other cases, I would say save yourself some money and use the Accel wire-wound yellow (or red) cables.

WATER PUMP

Any 5.0 Ford application could end up with one of two water pumps, these being the forward and reverse rotation items. Since we're never sure of the ultimate application of our project engines, life normally gets a lot simpler by dynoing with an electric water pump. Not only is this simpler for the test installation but, in most cases, there is horsepower to be had, as mechanical water pumps can absorb up to as much as 10 hp. We installed a CSI pump.

DYNO TIME

As one might expect, Keena-Levin was really looking forward to dynoing the finished project engine. We were now in the new UNCC Motorsports building, which has a dual-installation dyno cell. This was looking really good, and the gleaming 901 Superflow dyno was nearly ready to do battle with our engine. It all looked so photogenic, unfortunately one small, but highly significant detail, was to be our undoing here. The water

Accel's Universal-style ignition wire kits come with the boots on the plug end. Here, Mike Keena-Levin spends time cutting them to length and connecting them to the Performance Distributor's cap terminals for a neat installation.

T&L's dyno facility is just what you would expect of a Cup Car shop. It was spotlessly clean and made highly functional by attention to detail.

supply to the dyno cell proved to be too small, and a new, bigger system could not be installed in time to meet editorial deadlines. This caused something of a panic. Although we know plenty of people who have dynos, they are all busy doing race-car stuff (Charlotte is Cup Car country central). In addition to availability, we also needed a bigger-than-average dyno cell that was light and photo friendly (most are not).

Here is where what could have been a bleak situation reversed direction. Your author knew that his friend Lloyd McCleary, Busch engine builder (he has had as many as 25 engines in one Busch race) and boss at T&L had just the photo-friendly cell (three, to be precise) we were looking for, but is usually too heavily booked for an immediate test

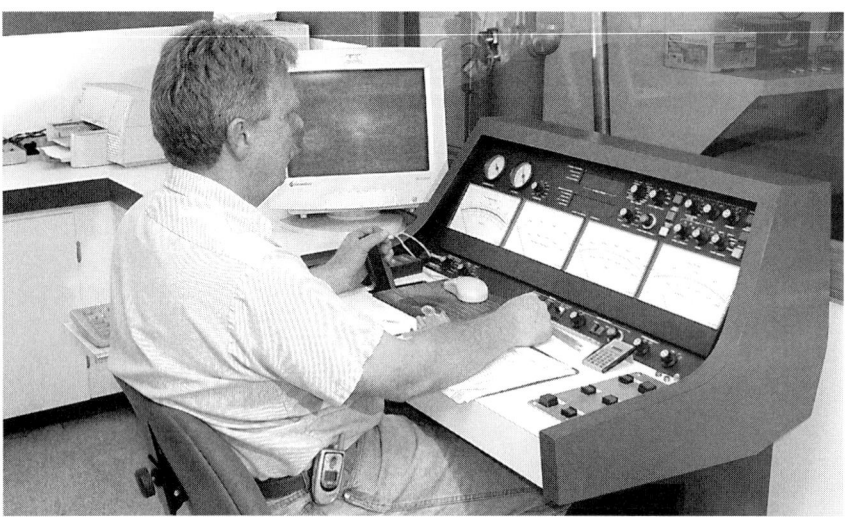

Here Lloyd McCleary, T&L's boss, runs the tests on our project 5.0. Once a few carb problems were ironed out, things moved along fast.

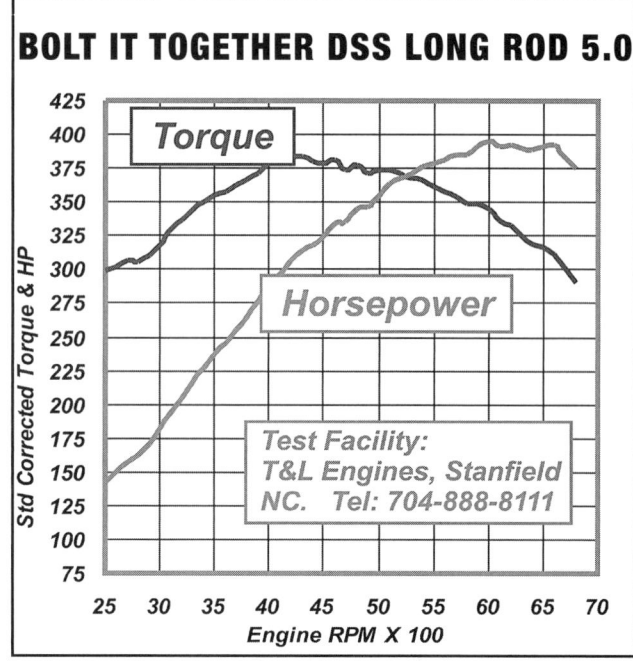

BOLT IT TOGETHER DSS LONG ROD 5.0

Test Facility:
T&L Engines, Stanfield
NC.   Tel: 704-888-8111

Even though we missed the 400hp mark by just a few horsepower, it can be seen that our D.S.S.-inspired bolt-it-together 5.0 dealt out a pretty mean power curve. Note how, for an out-of-the-box hydraulic valvetrain, it hangs on well at the top end.

session. Well, we got lucky. When we called, McCleary said to "be here Saturday morning at 11 a.m. We can run your engine 'til 4 p.m., but you must be out by 5." That may sound like a high-speed deal where it leaves us little time to do much, but McCleary's shop is geared to dyno a lot of engines. He sometimes has four engines through a cell in one day.

We arrived at T&L at 10:30 a.m., and by 11:00 we were wheeling the engine into the cell. About 20 minutes later, the engine was on and the Kooks headers were installed. The only problem we ran into was that the CSI electric pump was not instantly compatible with the hose arrangement on the dyno. Rather than spend time heading off to hunt down parts, we elected to swap out the CSI pump for a mechanical one McCleary had on hand. With that done, we were ready to go.

By 12:15 the engine was primed and ready to start. The BG Demon was given a half-dozen stabs to put fuel from the accelerator pump into an otherwise-dry manifold, and the starter was hit. After a couple of revolutions, the fuel hit the cylinders and the engine roared to life. Timing was set to about 5 degrees less than was expected to be finally used, and the break-in was carried out. Because of time constraints, it would have to be a short break-in, and we would have to forego an oil change to a more powerful synthetic like Mobil 1 or Joe Gibbs Racing's new synthetic race oil.

While the engine was breaking in, McCleary asked what the engine spec was. We told him what was contained within and that, with a few minor variations, it was essentially a D.S.S. build that we had done. "Well, those guys at D.S.S. must have a

good handle on things" he said. "We build a crate-motor spec for those race customers of ours who want a reliable street motor, and it is really close to that. If you guys have done a good job, you should see better than 370 lb-ft and 375 hp. If you are really on the ball, it should better 380 lb-ft and go right around 400 hp."

After only a 20-minute break-in, our first pull was done, and a huge miss instantly made itself apparent. Pulling the plugs revealed two cylinders were running unbelievably rich. This led us to suspect a blocked jet. A subsequent rebuild of the carb revealed debris probably acquired during our move into the new shop at UNCC, but we had no time to do a rebuild. McCleary had a number of Demon carbs on hand, so we grabbed the first-available photo-perfect 750 off the bench and bolted that dude on. Things got better, but it was still way too rich. We were about to go into it when McCleary remembered it was a customer's carb and had been set up a day or two before on a 550hp-or-so 383-inch engine. The next carb was one of McCleary's dyno specials (i.e., it never goes on a car). It was ugly from extensive use but had been on an engine that McCleary thought might not be too far off from what we wanted. Time was ticking away, and it was not until about 3:30 that we got to do the first real pull.

The carb was close but not spot on.

The engine made all the right noises on this pull and cranked out 378 hp along with 367 lb-ft—not bad for only 24 degrees of total timing. Now before you go thinking we should have set the timing at 34 degrees right off the bat, remember that one of the reasons the Edelbrock heads were used is that they have good swirl. This means a faster burn and, as such, less timing is needed to get the job done. Previous experience with heads having adequate swirl have shown they rarely need more than 29–30 degrees of total timing to maximize the horsepower. For the next pull, we ran 26 degrees, and output jumped by 12-some-odd horsepower and 14 lb-ft. This prompted going to 28 degrees total timing advance. The result was 385 lb-ft, 395 hp, and our deadline of 4 o'clock all at the same time.

Achieving 400 hp would have been nice, but no one was disappointed. Why? Because it was pretty much certain that another degree or so of timing would have netted a couple of horsepower as would the change to a synthetic oil. On top of that, a look at the plugs indicted some minor jetting changes would also have helped to an estimated tune of 3–4 hp. Given all these factors and a longer break-in, it looked fairly certain that this engine would have topped about 403 hp. As for its streetability, we found that the dyno, which was set up for high-rpm race engines, would not lug our 5.0 down lower than 2,200 rpm because of its strong, low-speed torque. At 2,200 it showed a steady state 289 lb-ft. That's more than a stock 5.0 peaks out at.

At the end of the day, we think we can say that not only did we get a good-looking engine, but also a cost-effective one that met all our needs for a true street machine as well as high performance on the track.

SOURCES

Accel
a Division of Mr. Gasket
10601 Memphis Ave. #12
Cleveland, OH 44144
216/688-8300
www.mrgasket.com

AED
2530 Willis Rd.
Richmond, VA 23237
804/271-9107
www.aedperformance.com

Barry Grant
1450 McDonald Rd.
Dahlonenga, GA 30533
706/864-8544
www.barrygrant.com

Calico Coatings
6400 Denver Industrial Park Rd.
Denver, NC 28037
704/483-2202
www.calicocoatings.com

Comp Cams
3406 Democrat Rd.
Memphis, TN 38118
901/795-2400
www.compcams.com

Crane Cams
530 Fentress Blvd.
Daytona Beach, FL 32114
386/252-1151
www.cranecams.com

CSI Performance Products
16936 County Road 252
McAlpin, FL 32062
800/226-1274
www.csiperformance.com

D.S.S. Racing
3550 Stern Ave.
St. Charles, IL 60174
630/587-1169
www.dssracing.com

Edelbrock Corp.
2700 California St.
Torrance, CA 90503
310/781-2222
www.edelbrock.com

Kook's Custom Headers
59 Cleveland Ave.
North Bayshore, NY 11706
866/586-5665
www.kookscustomheaders.com

Performance Distributors
2699 Barris Dr.
Memphis, TN 38132
901/396-5782
www.performancedistributors.com

T&L Engine Development
12303-A Renee Ford Rd.,
Stanfield, NC 28163
704/888-8111
www.tandlengines.com

Total Seal
11202 N. 24th Ave., Ste. 101
Phoenix, AZ 85029
602/678-4977
www.totalseal.com

Chapter 11
BIG
BOSS BLOCK

Major Engine Management With Ford Racing Performance Parts' (FFRP) Boss Block

Text by Steve Baur and Tom Naegele
Photos Courtesy of D.S.S. Racing and Ford Racing Performance Parts

Stolen Goods, our '93 Cobra project has created quite a stir in the Mustang community. Our almost unreal find of an SVT machine had a mere 1,300 miles on the clock, but the drivetrain was missing and we had to rebuild it and give it a new life. In the past few months, we got the car off blocks, and it's rolling again. Now it's time to get the heart of the snake beating—and since late-model Cobra Mustangs were essentially factory hot rods, we went back to the factory to find a suitable powerplant.

A COBRA BY TODAY'S STANDARDS

In bringing Stolen Goods back from hibernation, we wanted to achieve a performance level greater than that of the original '93 Cobra—or the R model, for that matter. When it comes to exceeding the factory engine output, that may be one of the easiest things to accomplish. Though the '93 Cobra and Cobra R powerplants were rated at a mild 235 hp, the same GT-40 components that SVT bolted on the engine were known to produce closer to 270 in independent tests. But even the latter mark is easily surpassed given the bountiful aftermarket of go-fast products.

If you look to 1995, the 351-powered Cobra R was still rated at only 300 hp. You have to move on to 2001 for more power, when the modular-powered Cobra R model's 5.4L, DOHC powerplant thumped out 385 hp and 385 lb-ft of torque. That's a stout number, but one we think Stolen Goods can eclipse, all while keeping the engine civilized and

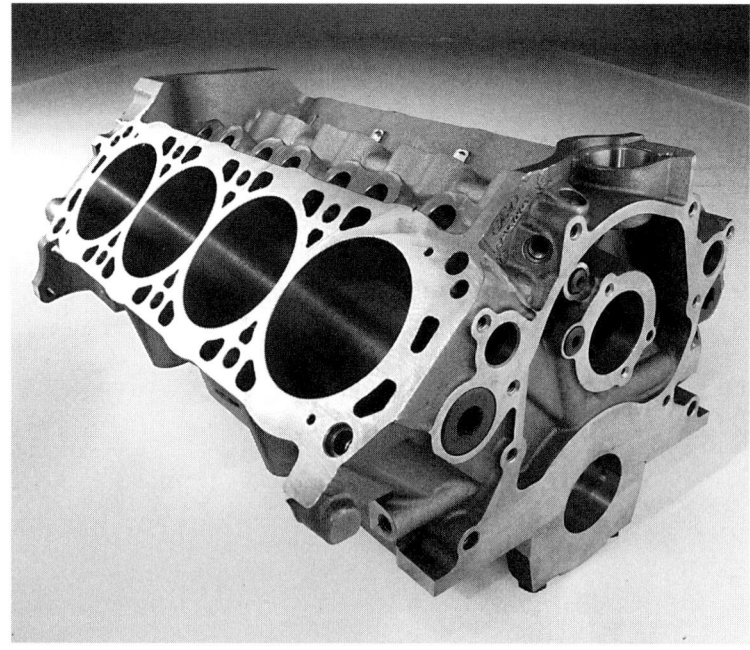

Ford Racing Performance Parts' Boss block (PN M-6010-BOSS302) retails for $1,759 and features numerous improvements over older designs. How much power do you want to throw at it? FRPP's not afraid.

The new Boss 302 block weighs about 32 pounds heavier than a Sportsman block, yet it's 16 pounds lighter than the R302 chunk of metal. It can be taken out to 363 ci and requires a minimum amount of work to sling the stroker assembly inside it.

Splayed main caps on journals 2, 3, and 4 will help keep the bottom end together under severe duress. The new Boss also employs 1/2-inch-diameter cap bolts for superior strength.

The Boss 302 utilizes common OD cam bearings or conversion bearings for use with most off-the-shelf cams. Stock cam journals, however, are all different sizes. Most cam companies offer common OD cams through special order, but FRPP came up with conversion bearings (PN M-6261-J351) to facilitate the use of common, off-the-shelf 302/351W grinds.

maintaining the broad torque curve 5.0L Mustangs are famous for.

From the beginning of this project, we never planned to do anything overly radical to our resident '93 Cobra, as we wanted a reliable and rock-solid platform to provide plenty of hours of high-performance fun without a lot of maintenance. Therefore we opted to go the naturally aspirated route with the engine. No blower belts to deal with, no trick fuel system needed, and no intercoolers or nitrous bottles to worry about, just raw horsepower and torque, all of the time. With that decision made, there were two ways to build the motor. We could design a high-revving 306, a stroked 302, or go with a big small-block Windsor.

To end up with a nice balance for open track, autocross, street, and strip driving, we set a goal of 400–425 flywheel horsepower. Another stipulation your author made was that Stolen Goods would have to make do with a 3.27, or at most, a 3.55 gear to keep the revs down on the highways. That pretty much ruled out the high-rev idea since the engine would spend far more time in the low and mid-rpm range. We also wanted to keep the stock hood, and while we've seen 351s fit under one, it usually came with great effort and mods to other items like the K-member to achieve this. So the Windsor was out.

Going with an 8.2-inch deck-height block would be the ideal choice, and since your author spends more time striving for deadlines than he does building engines, we turned to the Ford Racing Performance Parts catalog in search of an assembled short-block.

Part number M-6009-C347 is a 347ci stroker short-block assembly that uses a Sportsman two-bolt main block along with forged pistons, connecting rods, and crankshaft. That seemed like a pretty good choice, and it might be for many of our readers. We contacted Jesse Kershaw of FRPP to get his opinion and talked with him about our engine's intended use. That's when he offered us one of FRPP's new Boss engine blocks.

THE BOSS IS BACK

The new Boss block (PN M-6010-BOSS302) was designed with input from Ford's NASCAR engineers and from information gained during Ford's involvement in the ASA racing series. The Boss is set to be the basis for all of FRPP's crate engines in the near future, and it will be available as a bare or assembled short-block as well. At the time we went to print, assembled block production was still in its infancy. In fact, we received one of the

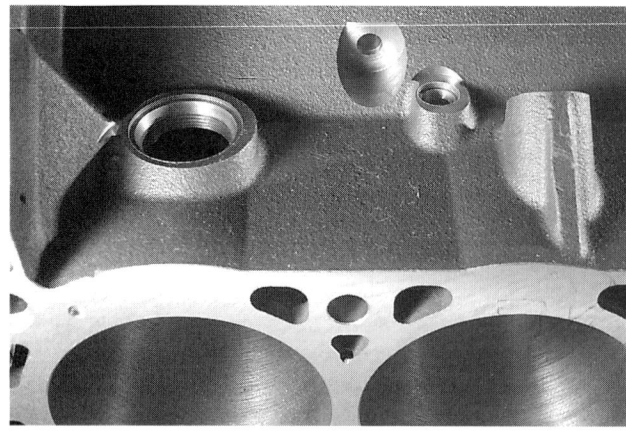

Screw-in freeze plugs and oil-galley plugs increase the strength of the block walls. These were utilized on the original Boss 302 block. Also present is the oil-dipstick hole, which allows use of the factory oil pan and dipstick.

The R302 block has screw-in galley plugs for oil but press-in plugs for the water. Dart blocks use press-in freeze plugs as well, unless machined by the builder for the screw-in–type pieces. The Boss 302 sets the standard with screw-in freeze and oil-galley plugs. The screw-in type of plugs adds more structural rigidity to the block and won't pop out in high-horsepower/high-stress applications.

first 150 blocks produced.

Its major features include four-bolt splayed caps on the number 2, 3, and 4 main crankshaft journals, screw-in, O-ringed freeze plugs and oil galley plugs, 1/2-inch main cap and head bolts, and a diesel-grade, nodular-iron casting that can be bored and stroked out to 363 ci. It is 16 pounds lighter than the R302 block it replaces and retails about $240 cheaper at $1,759.

Since we're not professional engine builders, we contacted Tom Naegele at D.S.S. Racing in St. Charles, Illinois, about constructing our snake's short-block. D.S.S. had assembled the 331ci Super Bullet bottom end in our project ProCharged Pony a few years ago, and the engine has handled everything we've thrown at it, so we knew it would be up to the task of building Stolen Goods' powerplant.

Naegele's plans for the Boss block includes using D.S.S. Racing's brand-new horizontal machining center to perform its CNC Level 20 blueprinting, giving us a state-of-the-art hunk of iron to use as the basis of our buildup. He also gave us his take on FRPP's new offering and expounded on some of its features as well as what D.S.S. plans to do to the new hunk of iron.

NAEGELE'S TAKE ON THE LEGEND REBORN

"For all practical purposes, FRPP's new Boss block is the R302 replacement," Naegele says. "Features include a four-bolt block similar to the R302 block. Main bolts are 351W 1/2-inch fasteners (R302 are 7/16), and in order to do this they had to move them outboard. This required FRPP to provide a new oil pump pickup due to space limitations. Keep in mind that there are no 302 blocks out there with 1/2-inch mains. We're going to install different main studs that are a little longer so that we can use the D.S.S. main support system.

FRPP added a boss and tapped hole for the clutch cross-shaft from early '60s and '70s cars. This wasn't offered on the Sportsman blocks. Now the old-school crowd can build a hot small-block, too.

"The head bolt holes are 1/2-inch pieces, like 351W and R302 blocks, but they're slightly deeper for better gasket retention and a lesser chance of ripping out threads. To utilize the extra length, you'll need different fasteners.

"The Boss' cylinders measure roughly 3.990 inches. The undersized bore needs to be finished bored and honed, just like the R block. The Boss' thick cylinders offer a comfortable 4.125-inch overbore capability—the R302 was stretched at 4.125 and often required sonic checking. It was more of a 4.100 piece.

"The new Boss block uses common outside diameter (OD) cam bearings or conversion bearings for use with most off-the-shelf cams. The R302 block was the first to come out with common OD bearings (Dart blocks use them also), and from a manufacturing standpoint, it's easier to bore the

Having consulted with engineers from its NASCAR teams, FRPP opted for siamese bores and provides coolant-cross overbleed holes to help remove trapped steam from the area.

cam bearing holes all the same size. Stock cam journals, however, are all different sizes. Most cam companies sell common OD cams through special order, but FRPP came up with conversion bearings to facilitate the use of common off-the-shelf 302/351W grinds. The conversion bearings are a bit expensive, though the cost has come down some from when they first came out.

"Another feature that FRPP added to the Boss block is its improved front and rear lifter-galley oiling feeds. The R block just oils the lifter galley from the rear and can starve the front lifters of lubrication. The Boss went the way of the Dart block, as it feeds the front and rear lifter galleys. In blocks with only rear oil feeds, FRPP noticed scuffing of the front lifter bores in some endurance applications. For the average Mustang enthusiast, this is not much of a concern, but it does add performance and value to the package.

"In the lifter valley, the Boss has finish align-honed mains and lifter bores—R blocks must be align-honed, as the bores are small out of the box. The Boss' big lifter bosses allow machining for offset or bigger lifters.

"One feature the Boss block shares with early Boss blocks are screw-in freeze plugs and oil galley plugs. The R block has screw-in galley plugs for oil but press-in plugs for the water. Dart blocks use press-in freeze plugs as well, unless machined by the builder for the screw-in type pieces.

"When you drop a stroker crankshaft in a stock 5.0 or R302 block, you need to machine the block for counterweight clearance, but FRPP has provided ample room with the Boss, and it has also done a nice job of cleaning up the crank case casting. Normally you could expect an hour to an

hour-and-a-half of grinding on an R302 block to fit the crank.

"The Boss has a shorter cylinder length, which makes rod bolt clearancing of the cylinders obsolete, but it may limit the amount of stroke and compression height you can run. The bigger you go on stroke, the more the piston protrudes out of the bottom of the cylinder. The Boss cylinders are approximately 0.400-inch shorter in length than a stock or R block.

"The deck heights on the Boss are 0.010-inch tall and will need to be equalized and decked just like the R block. This is fairly common among aftermarket blocks. The Boss features siamesed cylinders, and when FRPP was designing the block, the people in the NASCAR and endurance racing arenas that they talked with said the siamese design could overheat the head gaskets, so FRPP provided the Boss block with siamese cross-over coolant bleed holes to help remove trapped steam from trouble areas.

"In order to make the new Boss readily accepted by early Mustang and Ford enthusiasts, FRPP included a boss and tapped hole for the clutch cross-shaft from early '60s/'70s cars. This wasn't present on either the R block or the A4, which put a lot of people out when it came to building good motors for old cars with manual transmissions.

"Unlike most of the competition, the Boss block has a provision for the use of the factory dipstick, rather than requiring one to purchase a special oil pan–mounted one.

"The three center main caps are four-bolt and, more importantly, splayed. Splayed caps help keep main-cap walk under control, as they tie the cap

One feature that FRPP added to the Boss block is its improved front and rear lifter-galley oiling feeds. The R block oils the lifter galley only from the rear and can starve the front lifters of lubrication. In blocks with only rear oil feeds, FRPP noticed scuffing of the front lifter bores in some endurance applications, so the Boss feeds the front and rear lifter galleys.

The Boss has finish align-honed lifter bores—R blocks must be align-honed, as the bores are small out of the box. The Boss' big lifter bosses allow machining for offset or bigger lifters. They will, however, need to be machined down to use the factory lifter spider bar.

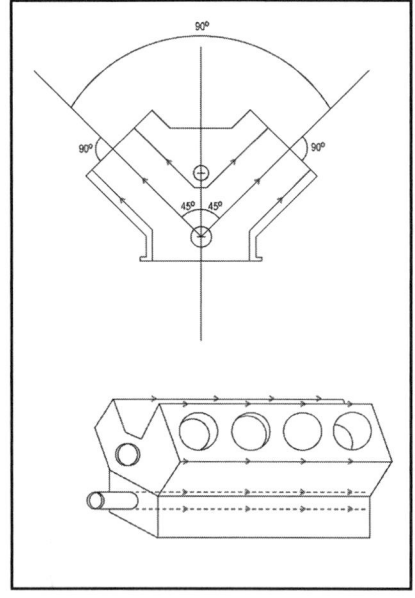

This graphic shows the main centerline and the engine's 90-degree V configuration. It's important that the cylinders are parallel to the centerline and perpendicular to the deck. If the cylinders are not parallel, they can be crooked, and compression can be different front to back.

laterally on an angle, which secures the cap, minimizing the sliding on the parting surfaces. Front and rear are still two-bolt pieces. The D.S.S. main support system that we'll use in this build will keep them from walking, and it also allows the use of a full-length billet-aluminum multilevel scraper/windage tray. Aluminum deadens the harmonics that make main caps walk. It does need longer main fasteners, and the main support comes with all of the necessary ARP custom hardware.

"The Boss block's big-bore nature is a win/win situation. While you can take it out to 363 ci (basically a 347 with a 4.125 bore size), this offers challenges when it comes to street-car longevity. The best engine combination is a result of the best compromise of all aspects of the engine, including rod ratio (rod length divided by stroke) and piston-ring placement. Engines that have poor rod ratios and use short piston designs tend to have the pistons rock in the bores, which causes excessive wear and added friction at a higher rpm. For racers, seasonal maintenance (rings and bearings) is not an issue, but it's not something someone with a street car is going to want to do. The short pistons also usually end up having the wristpins in the oil ring, and that can lead to oil control trouble if the proper ring package is not used.

"This is why many engine builders recommend the 331ci stroker assembly for street cars. The common 331 comes from a 3.250-inch stroke, a 4.030-inch bore size, and a 5.315-inch connecting rod. This combination results in a 1.250-inch compression height, which provides proper spacing for the piston rings. It is far superior than the 1.090-inch height found in common 347ci motors. The better rod ratio and taller piston results in better ring placement and spacing. Subsequently the piston

Since the new Boss 302 block uses larger 1/2-inch main-cap bolts, a different oil-pump pickup was needed, and FRPP has you hooked up. Part number M-6622-Boss302 is what you need to fit the typical Fox stamped-steel pan.

rocks less in the bore and promotes greater ring-and-piston life. The taller piston prevents the wristpin from intersecting the oil ring and a better 1.63 rod ratio is achieved.

"The common 347ci engine, with its 3.400 stroke, 5.4-inch rod, 4.030 bore, and 1.090 compression height (the distance from the wristpin center to the top of the piston) makes for the same inefficient piston/ring/rod ratio design and results in a 1.58 rod ratio just like the 363.

"With the Boss block's big-bore capability, we can obtain the 347 ci along with a 1.63 rod ratio. Using a 3.250-inch stroke, a 5.315-inch connect-ing rod, and a 4.125-inch piston, we arrive at 347 ci with the 1.63 rod ratio—a 347 done the right way."

D.S.S. BLOCK BLUEPRINTING

"People often toss around the term blueprinting, but few understand it," Naegele continues. "Our new horizontal machining center was purchased because of the need and desire to CNC blueprint a block more accurately than was previously possible. Blueprinting is remachining the critical dimensions and geometry to correct it based on factory specs.

"The crankshaft main centerline is pretty much ground zero, and cylinders should be 90 degrees apart and equidistant from the crank and camshaft centerlines—each bank should be 45 degrees from the centerline. Another important part of blueprinting is ensuring the deck surfaces are perpendicular to the cylinder bore, which should be directly located over the crankshaft and parallel to the 90-degree centerline.

"From the factory, OEMs just need blocks to have similar compression between cylinders. The closer you get to a perfectly machined product, the better it will run and the longer it will last. Because

	THE ORIGINAL 1969-70 Boss 302	GOOD Sportsman 302	BETTER R302 Race Block	THE BOSS Ford Racing BOSS 302
Part Number	N/A	M-6010-B50	M-6010-R302	M-6010-BOSS302
Main Caps	4-bolt cast iron (2, 3, 4)	2-bolt cast iron	4-bolt nodular iron machined splayed (2, 3, 4)	4-bolt nodular iron machined splayed (2, 3, 4)
Siamese bore	No	No	Yes without cross drilling for cooling	Yes with engineered cross drilling
Freeze plugs	Screw-In tapered pipe thread	Press in	Press in	Screw-In O-ring sealed straight thread
Material	Cast Iron	Cast Iron	Cast Iron	Diesel grade heat treated cast iron (41,000psi tensile strength)
Head bolts	7/16 "	7/16 "	1/2 "	1/2 "
Recommended Max. Bore	4.030 "	4.030 "	4.125 "	4.125 "
Front oil crossover for lifter galley	No	No	No	Yes
Main bolts	7/16 " inner (1-5), 3/8 " outer (2, 3, 4)	7/16 "	1/2 " inner (1-5), 3/8 " outer (2, 3, 4)	1/2 " inner (1-5), 3/8 " outer (2, 3, 4)
Oil galley plugs	Pipe thread	Pipe thread and press in	Pipe thread	Screw-In O-ring sealed straight thread
Hydraulic roller compatible	No	Yes	Yes	Yes
Clutch cross shaft pivot hole	Yes	No	Yes	Yes
Rear main seal	2-piece	1-piece	1-piece	1-piece
CID capacity	347	347	363	363
Weight	143 lbs	133 lbs	181 lbs	165 lbs
MSRP	N/A	$1,050	$1,999	$1,759

of the slack in factory tolerances, compression from bank to bank will be different. D.S.S. has seen a minimum of 0.005 inch difference, and as bad as 0.025. The OEM engines just need to run and last through the warranty period; however, setting deck height and ensuring it is equidistant from the crank centerline is extremely important to get the most out of your motor.

"Many places use the Sunnen CV616 or CK10, as they're the industry standards and are designed to hone 0.030 over without boring the cylinder. If you do not bore the cylinder and then hone it, you are not correcting the cylinder; you're just averaging the wear and not straightening out the cylinder. You are also moving the cylinder over in the direction of wear. It'll run fine, but that's the difference between average and truly blueprinted.

"Honing cylinders with torque plates is also an important part of blueprinting. A lot of CNC equipment fixtures blueprint off the oil-pan rail, which is how Ford machines its blocks, but it's not the main bore and not a good reference point. The fixture that D.S.S. manufactured to do its CNC work is the only one of its kind, and it allows us to blueprint off of the mains instead of the oil-pan rail.

SOURCES

D.S.S. Racing
3550 Stern Ave.
St. Charles, IL 60174
630/587-1169
www.dssracing.com

Ford Racing Performance Parts
15021 S. Commerce Dr., Ste. 200
Dearborn, MI 48120
800/367-3788
www.fordracingparts.com

"Oddly enough most aftermarket automotive CNC and conventional machines mimic the same OEM set up off of the oil pan rail. Also, most bore off of the OEM cylinder and true the decks using a bubble level. For Joe Average, that's OK, but it should be perpendicular. Most boring machines go off of the deck surface, so if it isn't perpendicular, then cylinder bores won't be exactly 90 degrees from the crank.

"All of Level 10 and 20 CNC blocks are blueprinted off of the mains, something D.S.S. has been doing for the last five years. We invested heavily in this method, and though it's more profitable to do it the old-fashioned way, this is the right way to do it. Holding a block rigid enough to machine within a tenth or two vertically off of the main centerline is not easy (or cheap), but the countless hours of CAD design and machining paid off. This is by far the most accurate, rigid block fixture and machine we have ever seen.

"Why is that so important? Here's an example. Camshafts are ground with lobes 90 degrees apart, but what if your block was 89 or 91 degrees? What if your compression ratio was three or four tenths of a point different side to side or even front to back? Of course it will run, but it won't be optimized, and in a high-performance application, getting the most from what you have is of the utmost importance. Our Level 20 CNC blueprinting on the Boss block also includes the required lifter boss relieving so the factor lifter retaining hardware can be used.

"Twenty-five years of racing and honing thousands of blocks has taught us plenty about the importance of ring seal and cylinder prep. The three-step honing process we developed is done on all of our Level 10 and 20 blocks. These steps are pivotal to our customers' success."

Here is our Boss 302 block mounted in the D.S.S. fixture. As you can see from the picture, it allows D.S.S. to machine off of the main cap centerline for extremely accurate results.

The new Boss 302 block is getting its lifter valley machined for the factory lifter retaining hardware

We'll take the rough 3.99-inch cylinders out to a finished 4.125 bore size. This, combined with our 3.250-inch stroke and 5.315-inch connecting rod, will take the Boss to 347 ci.

Chapter 12

MODULAR MAGIC

Livernois Motorsports Shares the Secrets of Building a Killer 4.6L Short-Block

Text and Photos by Samuel James

The popularity of 4.6 modular V-8s has grown exponentially over the last couple of years. While the overhead cam family of engines has not yet eclipsed the popularity or the levels of performance of the small-block pushrod family of 5.0s and 351s, the gaps are getting narrower every day.

In many ways, the mod motors far surpass the small-blocks we like to call the Windsor family—block design, strength, and cylinder head technology are way ahead of anything the factory produced, especially in Four-Valve Cobra trim. The 4.6 (and its 5.4 cousin) have proven to be rugged and reliable, as well as easily adapted to a wide range of horsepower and performance levels.

Livernois Motorsports is a performance engine shop and parts retailer specializing in Ford products in general and the 4.6 modular in particular, although the company also offers a broad line of other domestic V-8 performance engines as well. The 36,000 square-foot shop just outside of Detroit is a Ford lover's dream, with non-stop trick parts and engine-building secrets, including complete machine shop with full CNC capabilities, hard-core racing parts, installation center, and on-site dyno tuning. *Muscle Mustangs & Fast Fords* paid a visit to see if the Livernois staff would whisper some of their secrets in our ear.

We talked with Mike Schropp (shown above), head of the Modular Engine Development Program at Livernois. Schropp has many years of experience in designing and testing modular engines, and he knows just what to do to make them live through the most grueling punishment, from killer street motors to extreme race pieces. We asked him to tell us what's involved in building a performance modular short-block. He was nice enough to explain to us much of what goes into building a Livernois modular short-block.

Schropp says Livernois offers modular short-blocks in a variety of configurations, including stroker and non-stroker, depending on the needs of the customer. The company will also custom-build a short-block or complete engine to the customer's specs if desired, including Two-, Three-, and Four-Valve configurations, with an aluminum or cast-iron block.

Work areas are entirely separate from the assembly area in order to contain machining debris.

The cast-iron modular block has excellent rigidity and stability. When line-honing aluminum modular blocks, Livernois' Mike Schropp installs and torques the oil pump as well as the torque plates so that the block is stressed exactly the same way it will be after assembly.

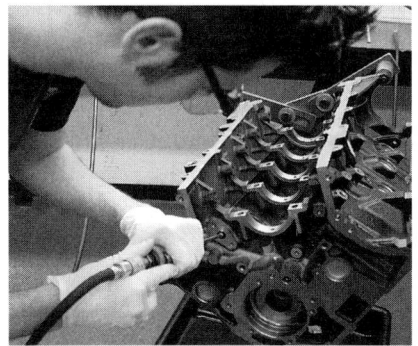

Oil passages require special attention. Schropp has found that he is able to radius and deburr them to improve oil flow and engine longevity.

Likewise, Schropp has found that chamfering the oil holes in the crankshaft will enhance oil flow to the rod and main bearings, critical in any performance application.

While not needed for line-honing the main bearing bores, deck plates are a must for honing cylinder bores in order to replicate the stresses that will exist with the heads torqued into place.

Here he deburrs the bottom of the cylinders. These modular cast-iron blocks are made very well and require a minimal amount of deburring.

All crankshaft journals are micro-polished to minimize friction and maximize precision.

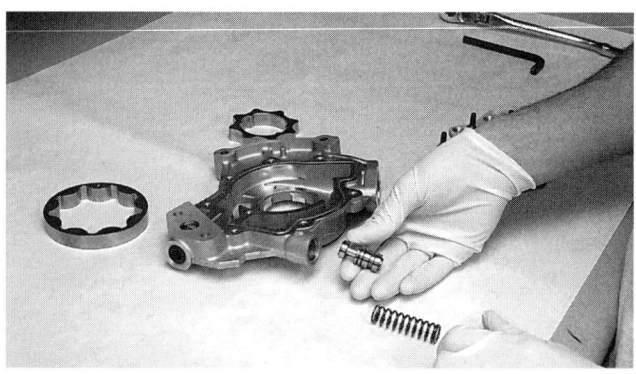

This is the oil pump plunger that Schropp has, on several occasions, found installed backward from the factory. This is one of the reasons he doesn't trust factory assembly, but rather takes responsibility for every component in the engine by inspecting and verifying each part.

While many of the short- and long-blocks sold by Livernois are boosted courtesy of a supercharger, turbocharger, or nitrous, the machining and assembly procedures are pretty much the same regardless; the biggest differences are found in the choice of pistons, rods, and crankshaft. For this story, Schropp built a typical short-block that would be used in a Two-Valve engine with a centrifugal supercharger, which has proven to be a popular configuration, making as much as 600 rwhp with excellent driveability and reliability.

So here, then, are some of the secrets that go into the building of a Livernois modular short-block.

PARTS ARE PARTS ARE PARTS—NOT!

Thinking that all parts are the same is like thinking the human body is just 20 bucks worth of chemicals—it's not just the raw materials, it's the magic of the mixture. Our Livernois guru explains that 4.6 modular blocks have been offered in aluminum or cast iron, with the cast-iron blocks made in Windsor (Ontario) as well as in Romeo (Michigan). The Windsor engine uses dowel pins to locate the main bearing caps, while the Romeo block uses indented-hex jack screws. All aluminum blocks use the jack screw arrangement except for the latest aluminum castings which do not use jack screws at all.

For cast-iron applications, Livernois uses new Romeo blocks from Ford. These bare blocks are supplied with main caps and bolts, are of consistently high quality, and are reasonably priced.

For customers who want to go with aluminum blocks, Livernois will machine and build the customer's bare block core; new aluminum blocks are the later Windsor style. The aluminum blocks use cast-iron liners, and can be safely bored 0.020 inch over. Schropp explains that they can install a

Here's the sophisticated profilometer used at Livernois to measure surface finishes, including cylinder bore finish. Note that it is used with deck plates installed for added precision and accuracy.

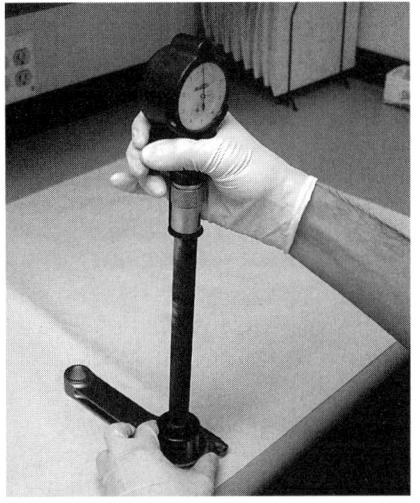

A dial bore gauge is used to measure the exact inside diameter of the connecting rod/bearing assembly. After mic'ing the corresponding crankshaft journal, a little simple arithmetic reveals the precise bearing clearance.

Prior to measuring piston ring end gap, Schropp installs each ring square, and true to the bore, at exactly the same depth.

Even individual piston rings are labeled during the blueprinting process to assure that the location where they're measured is the same location they'll end up in after final assembly.

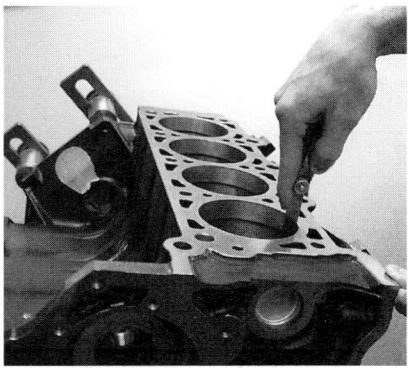

Top ring end gap is specified at 0.020 inch +/- 0; the second ring is specified at 0.022 inch +/- 0. There's no room for deviation here; end gap is entirely at the control of the engine builder and his ring file.

Here's the fish-scale method Schropp uses to measure the drag of the rings in the cylinder bores. Any deviation from his knowledge base serves as a trigger to dig more deeply for the cause.

big-bore liner kit, which takes the bore out 0.150 to 3.700 inches, resulting in displacement of a nominal 302 ci. Depending on the customer's needs, however, it's usually more cost-effective to go to a stroker crank instead of installing eight new cylinder liners.

Most of Livernois' crankshafts come from Kellogg, a leading supplier of OE cranks for Ford and other car makers, and for the performance aftermarket as well, and customers can specify forged steel or billet cranks, in standard or stroker configurations. They do not use cast cranks at all. Schropp also says that Livernois has sophisticated five-axis in-house CNC capabilities and does custom work on different cranks.

Livernois uses Manley H-beam connecting rods in almost all modular engines, for standard or stroker applications, however some get pro-series or billet rods for extreme applications. Schropp points out that Manley's stroker rods for the 4.6 are sold exclusively through Livernois. He likes the Manley rods because of their strength and light weight. All use full-floating wrist-pins, and are fitted with upgraded ARP 2000 rod bolts.

Schropp prefers Clevite H-series performance bearings, and to minimize horsepower-robbing friction, Livernois is now starting to use Clevite's recently introduced TriArmor-coated connecting rod bearings. This technology uses a moly-graphite coating just 0.0003-inch thick applied at the factory, to Clevite performance bearings. This super-slick coating reduces power-robbing friction, while providing protection during cold startup. It is designed to be sacrificial, such that, even when it starts to wear off after many miles, there's still a complete, unmodified performance bearing underneath to provide many more miles of use. Main bearings, which are not subjected to the high stresses inherent in reciprocating connecting rods, are standard H-series Clevite performance bearings. While these have proven to be durable, rumor has it Clevite is developing TriArmor-coated main bearings for the 4.6.

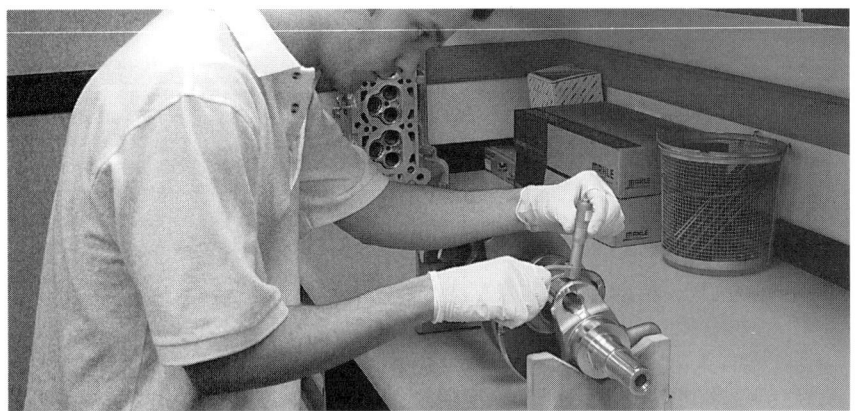

Each crankshaft journal is mic'ed as a double-check to assure precise clearances, and help maintain Schropp's reputation as never having had a comeback on any engine he's ever built.

Here Schropp measures connecting rod side clearance.

After the assembled main bearings are measured, our Livernois expert notes the location of each bearing shell so that it goes back into the same location during final assembly.

Crankshaft endplay is a critical tolerance. Too little will lead to binding and premature failure of the thrust washers. Schropp select-fits or hand-laps thrust washers as necessary to achieve the desired endplay.

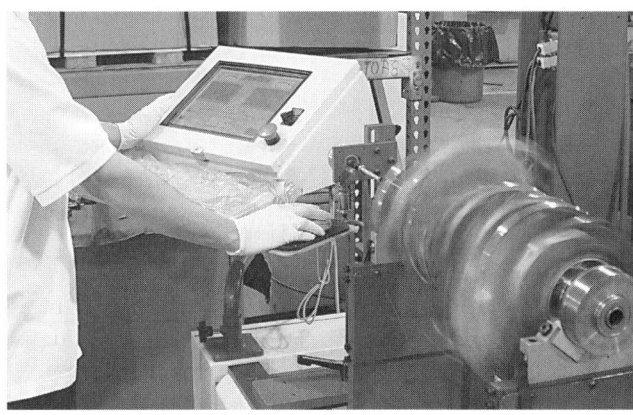

High-speed balancing of the rotating assembly helps eliminate power-robbing harmonic vibrations that can appear at high engine speeds.

While the use of a torque wrench is adequate for many fasteners in an engine, measuring bolt stretch is more accurate, and Schropp prefers this method for the super-critical connecting rod bolts.

Livernois sources pistons exclusively from Mahle. Some are off-the-shelf parts, while others are built to Livernois' designs. Differences lie in the top of the piston, which may be flat top or have one of about a half-dozen dish configurations, as well as stroker and standard configurations, depending on the desired compression ratio, and if/how the engine may be boosted. All of these Mahle pistons are forged from M124S (4032) aluminum alloy, and are phosphate coated to protect against ring micro-welding and pin galling, especially during initial startup and break-in. The combination of low expansion alloy for tighter piston-to-wall fit allows better piston control, as well as improved ring seal and blow-by control.

Rings for this engine are by Perfect Circle. Top rings are barrel-faced ductile iron with plasma-moly facing and no inside bevel, meaning they are non-twisted; second compression rings are reverse-twist gray cast iron. Oil rings are Perfect Circle's popular three-piece CP-20 design commonly used in performance engines. Schropp prefers standard tension oil rings for these engines, although Perfect Circle also offers them in reduced tension configurations to suit the preferences of the individual engine builder. These rings are supplied by Perfect Circle in ready-to-install sizes or +0.005 inch, which Schropp prefers in order to allow him to file-fit the ring end gaps.

Oil pumps are stock Cobra/GT units as supplied by Ford. Almost all engines he builds are wet sump, and most oil pans are stock Ford, although he'll use pans from Moroso or Canton if the customers ask or if necessary for a particular application. Schropp notes that it's important to consider the entire lubrication system as a whole, since every clearance in the oil system affects pressure and flow. His years of experience with this engine family have taught him to consider every element of the oiling system

Schropp cleans each cylinder bore repeatedly until the lint-free towel comes out perfectly clean. Even microscopic grit can prevent proper ring seating, which would lead to reduced power and longevity.

SEALING CYLINDER PRESSURES
The Ring's the Thing

Of course, the role of piston rings is to seal combustion pressures, right? Well, yes and no.

Certainly, containing cylinder pressures is one role of the piston ring pack, but Clevite's Bill McKnight, whose company also supplies Perfect Circle piston rings, explains that it's not nearly that simple.

"A typical performance piston carries two 'compression' rings and an oil ring assembly," explains McKnight. "But there are subtleties some engine builders aren't aware of. For instance, many people don't realize that 80 percent of the role of the second ring is actually oil control, not compression control.

"Furthermore, the spring tension of the ring against the cylinder wall is not at all sufficient to contain combustion-chamber pressures. It is essential for the ring to be designed such that combustion pressures are routed in behind the ring to increase outward pressure against the cylinder wall.

"The Perfect Circle top rings used in this Livernois buildup are of a barrel-face, non-twist design. P.C. engineers have found this design to be their most effective in the modular 4.6 Ford engine."

as it relates to the engine's power and expected usage. He carefully plans every phase of the lubrication system, from the oil pump pickup to the pump clearances and relief valve spring pressures all the way to main and rod clearances as well as rod side clearances and valvetrain clearances. His hundreds of dyno sessions with modular engines, along with all of his builds and end-of-season teardowns, have taught him why naturally aspirated road race engines should be engineered differently from turbocharged or supercharged drag race engines.

BLUEPRINTING IS KEY

According to Schropp, blueprinting at Livernois involves preparation of all internal engine parts and measuring and dialing-in all clearances. This is the tedious part; once all of this is done, as you'll see, final assembly is pretty straightforward. As components are blueprinted, each part is num-bered or otherwise identified for positioning during final assembly.

Because he has built so many of these modular engines in different configurations, Schropp has learned which clearances yield optimum power and durability in different applications. He is adamant that no two engines are alike, so the clearances used for this engine are only for this particular engine, and will be different in engines that will be built with other parameters and for other circumstances. Schropp's vast experience has enabled him to compile an extensive body of knowledge far beyond that of the casual engine builder.

Blueprinting begins with preparation of the engine block, which receives a complete deburring. Schropp takes pains to radius all edges and corners and to remove all potential stress risers, especially at the parting lines in the casting. This deburring procedure is extended to the oil galleries, including the smoothing of corners and radii within the oil galleries. All oil gallery plugs are removed and tapped for screw-in plugs. In addition, cylinder bores are chamfered, as are all sharp edges that may bear against moving parts.

As for the deck of the block, our Livernois expert explains the factory deck surface on new iron blocks is generally just fine for most applications. For all-out race engines, Schropp will "parallel-deck" the block to assure identical deck height for both banks.

Special attention is given to the main bearing bores to assure perfect main bearing fit and alignment. New ARP main studs are installed, and the nuts are torqued and loosened a total of three times to set the threads of the stud against the threads of the nuts to assure repeatable torque values and tension.

Main bearing housing bores typically have some taper and variation in bore size and roundness from the factory. Schropp installs the Clevite main bearings and torques them to spec in order to establish a starting point for bearing clearances. Following disassembly, main caps are cut for a nominal housing bore size of 2.850 inches, and the block and caps are align-honed to the correct size. The housing is checked throughout the honing process to assure the housing bores are straight, free of taper, and perfectly round. Then, after cleaning, main bearings are reinstalled and final clearances checked to assure that they are 0.0029 inch +/- 0.0001 inch.

Cylinder bores are honed on a Sunnen CV-616. Schropp fits the pistons to individual bores, with a piston-to-wall clearance of 0.0035 inch +/- 0.0001 inch. Cylinder wall finish is confirmed using an unusually sophisticated profilometer. Most profilometers can read about four parameters including roughness average (Ra). Livernois' computerized profilometer can measure more than a dozen parameters, including plateau height, total depth, and so on. This is especially helpful in determining oil control, which is related not only to ring seating, but also to the amount of oil held in the "nooks and crannies" of the cylinder wall. This is confirmed by Perfect Circle piston ring engineers, who have observed more precise oil control in engines built with the benefit of such a profilometer.

The crankshafts that Livernois receives from Kellogg are balanced to +/− 3 grams. Journals are polished 15–20 seconds each, with a 2,500-grit belt in order to achieve a chrome-like finish. Oil holes are chamfered, and the entire crankshaft is deburred, including all holes, edges, and any potential stress risers. The crank is carefully washed at this point, although this is not the last cleaning the crank will see before final assembly.

As for the connecting rods, Schropp hones the small end of the rod to achieve 0.0010-inch pin-to-bore clearance. After sizing the big end, he mic's the crank journals and select-fits bearing shells to achieve 0.0025 inch clearance +/- 0.0001 inch. The Clevite TriArmor-coated rod bearings are available in standard size as well as "X," which designates 0.0005 inch for each shell, or a total of 0.001 inch oversize, allowing for more precise select-fitting. Clevite is careful not to coat the parting surfaces of their TriArmor rod bearings so as to not disturb crush, which is so critical to bearing performance. Rods are then torqued into place in pairs, with the feeler gauge left carefully positioned between adjoining rods in order to prevent the rods and bearings from cocking during tightening, which would affect rod bearing clearances and alignment, and also side clearance. After tightening, rod side clearance is checked, and rods are hand-sanded as necessary to achieve 0.011 inch side clearance +/- 0.0005 inch.

Out of the box, the Mahle pistons are almost assembly-ready. They are already pin-fitted, and usually require no deburring or radiusing. Typically the pistons are balanced +/- 0.5 gram from the factory.

All piston rings end gaps are file-fitted. This engine will have a top-end ring gap of 0.020 inch, +/- 0 (don't you just love that? Plus-or-minus 0? Now that's accuracy!). Second ring end gap is

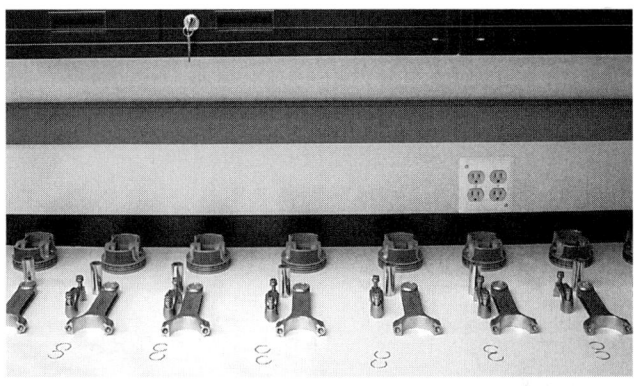

All internal parts are laid out for final assembly. Schropp notes that the discipline of laying all the parts in order becomes part of a routine that assures consistent, reliable engine builds. It also helps any inconsistencies or irregularities to leap out at him as exceptions to his tried-and-true practices.

specified at 0.022 inch, again +/− 0. Side clearance for the compression rings is 0.001–0.0012 inch. The three-piece Perfect Circle oil ring (two rails and an expander) is select-fit to achieve rail-end gap of 0.020 inch +/- 0.002-inch. File-fitting is done with a high-end Goodson 110-volt ring file that incorporates an adjustable-bore plate and dial indicator. After file-fitting, rings are hand-filed to deburr and radius, then are final washed.

Schropp strikes a balance between oil control and reduced friction by installing the oil ring pack on a piston/rod assembly and using a fish scale to precisely measure the force needed to pull the piston (with oil rings) through a lightly-oiled cylinder. This allows him to fine-tune oil ring tension for each engine. For example, a dry sump engine with vacuum pump can benefit from the reduced friction of a low-tension oil ring pack, while a boosted engine is better served by the improved oil control of a higher oil ring tension. Schropp's wealth of experience allows him to combine the somewhat low-tech fish scale with the sophistication of a profilometer to get an exact cylinder-wall finish and a perfectly tailored ring package for the precise needs of the engine at hand.

Although the oil pumps are brand-new from Ford, they are disassembled and inspected for casting flash. Cover and rotor clearances are checked, and all components are deburred as necessary. Also, Schropp is careful to remove and inspect the pressure valves, since, on several occasions, he and his people have seen these valves installed backward from the factory. Schropp prefers to fill each pump with Clevite Bearing Guard lubricant before reassembly. Bolts receive a coating of Loctite and are torqued into place.

Here's the bare block with bearings installed and lubed, ready for the crankshaft.

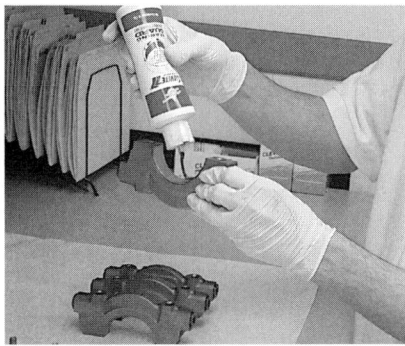

Even though the Clevite TriArmor bearings have a protective coating applied during manufacture, Schropp adds extra insurance with Clevite assembly lube.

Since ARP fasteners are used in these engines, Schropp uses the special ARP moly lube for all fasteners. ARP provides different torque values for their fasteners depending on whether moly lube is used, since it is so effective in reducing friction.

As with the rod and main bearings, thrust washers are lubricated with engine assembly lube for protection during start-up.

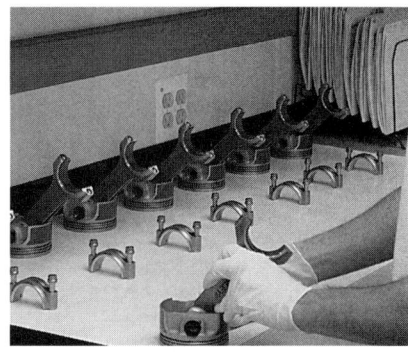

The fit of the piston pin is critical, and Schropp knows the feel of a perfect fit.

While we haven't mentioned it at each step of the blueprinting process, Schropp insists we point out that every component is washed or cleaned several times during the course of blueprinting. This assures that measurements are not compromised, and also protects against cross-contamination from one part to another. It also precludes occurrences like embedding grit into soft bearing surfaces.

FINAL ASSEMBLY

The recurring theme throughout final assembly is cleaning, cleaning, cleaning. While it may sound repetitive, so is the cleaning process itself. Schropp prides himself on never having had a comeback due to dirt or contamination (or any other reason, for that matter), and he intends to keep it that way.

With the block, for example, Schropp cleans all oil galleries with rifle brushes, then it goes into a jet wash tank for 15–20 minutes. Then he sprays it down with clean water and air dries it before transporting it to the super-clean assembly area. Upon arriving in the engine build area, all oil holes are sighted and checked for zero contaminants of foreign debris or dirt. Then threaded oil gallery plugs are installed.

The previously numbered upper main bearing shells are wiped clean and installed. The freshly polished crank goes to the wash tank, is dried with compressed air, and then receives a final visual inspection, especially in all the oil passages in the crankshaft. Schropp prefers to lube all bearing surfaces with Clevite bearing lube; he likes this particular product because it is very viscous and stays in place throughout the assembly process and during storage until the engine is fired.

Our engine expert lubes the thrust side of the thrust washers and places them into position, then installs the main bearing caps with the pre-numbered and well-lubed bearings in place. Even though the moly-graphite coating on the Clevite TriArmor bearings provides protection during initial start-up, Schropp applies a liberal amount of bearing lube as added insurance. After the main caps are in place, only then does he screw in the ARP main studs. Installing the studs after the crank and main caps are in place eliminates the possibility of nicking the crankshaft or bearings during assembly.

Main studs and washers are coated with the supplied ARP moly lube prior to assembly. Studs go in hand-tight only—never more, since tightening these studs can actually distort the main bearing housing bores. Schropp then begins tightening from the center out, to half the final torque spec of 65 ft-lb. He notes that some shops miss the important step of "setting the thrust"—assuring that the thrust main and thrust washers are perfectly perpendicular to the centerline of the crank.

He accomplishes this by loosening the No. 5 main cap nuts and just snugging them back up (since this

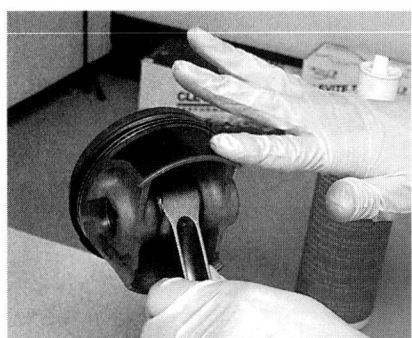

Schropp carefully applies lubricant to piston skirts to facilitate assembly, but also to minimize friction during initial start-up. This protects ring-to-wall contact and assures proper ring break-in during this oh-so-critical initial start-up period.

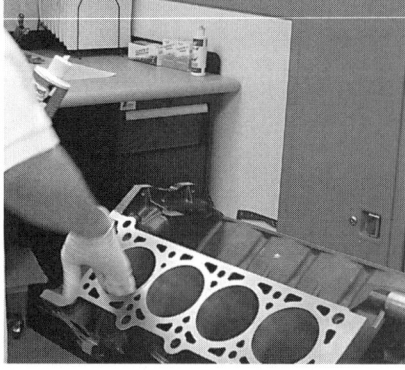

Cylinder bores receive a final inspection before Schropp installs the piston/rod assemblies.

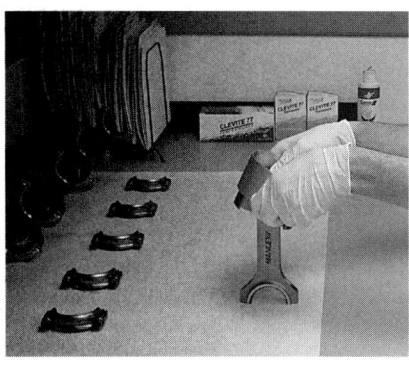

Schropp prefers a tapered ring compressor to the less expensive but more commonly used band-type. The tapered design does not have the "seam" that can scratch the rings and compromise their performance.

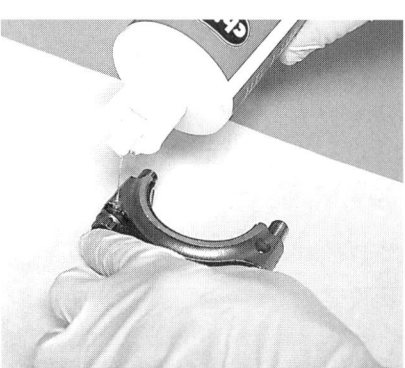

He lubes each rod bolt to eliminate the possibility of any binding or other interruption of bolt torque during the tightening process. The bolt must be free in order to achieve the proper stretch.

A "persuader" is used only to move the pistons through the ring compressor. Once the rings are all within the cylinder, Schropp prefers to pull the rods the rest of the way into position by hand in order to assess the friction factor.

is where the thrust is taken on the modular engine), then carefully prying the crankshaft forward and applying one carefully calibrated blow with a dead-blow hammer to the rear of the crankshaft. This moves the No. 5 main cap ever so slightly within the free play on the studs, so that the thrust washer is perfectly flat against the thrust surface of the crank. He tightens the No. 5 nuts back to 32.5 ft-lb, then goes back and tightens all main bearing nuts, in sequence, to full spec.

The next step is installation of the 10 main cap jack screws, tightening them from the center out, first to 45 in-lb, then to final torque of 90 in-lb. Then new side bolts go in, torqued to 20 ft-lb. These side bolts are of a torque-to-yield design, which means they must never be reused. So Livernois uses two sets on every engine they build—one set for line-honing, and a new set for final assembly.

Now that all main bearing fasteners are tightened

properly, Schropp checks endplay with a dial indicator. He measures in several locations, looking for 0.006 inch, +/- 0.0005 inch. Any variance can be eliminated by select-fitting or hand-lapping the thrust washers.

Final cleaning of the cylinder bores is critical to ring seating and life. Schropp cleans the bores several times with lacquer thinner, then WD-40, and then with lacquer thinner again. This process alone can take more than a half hour, and is only complete when the lint-free cloth comes out of the bore as clean as when it went in. Next is assembly of the pistons, pins, and locks to the rods, with Clevite engine lube on the pins.

As with the bearings, the piston rings were numbered during the blueprinting process, and installation of the Perfect Circle rings on the pistons with a ring expander is fairly straightforward. But Schropp does take great pains to offset the ring end gaps, for two important, yet subtle, reasons. It is widely accepted, and confirmed by Perfect Circle engineers, that rings rotate when the engine is running. So why bother to offset the end gaps during assembly?

With the ring end gaps offset, there's slightly more cylinder pressure during initial start-up and less chance you will end up with all the gaps lined up, which could happen if you started with them all lined up. Offsetting the end gaps also helps promote ring seating, essential for ring life.

Schropp lubes pistons, rings, and cylinder walls with Childs & Albert SAE 30W oil and, with the help of a ring compressor, slips all eight piston/rod assemblies into their bores with the upper bearing shells in place. He prefers to pull the rods into

All tightening is done by hand, so Schropp can feel for any burrs or other irregularities. It's all part of the repetitive process—like an airline pilot's check list—that makes any inconsistencies jump out.

Experience has taught him that final tightening of the rod bolts is best done with the feeler gauge in place in order to prevent cocking of the rod.

position on the crank by hand to get a feel for the fit and friction, pointing out that, if you have to tap the piston down hard enough to dislodge the upper bearing shell, something is amiss.

Schropp has found the use of a bolt stretch gauge to be the most accurate means of tightening rod bolts. He tightens each bolt, lubed with SAE 30 engine oil, to spec (65 ft-lb) and then loosens and retightens each one a total of three times before final tightening, in order to assure the bolt is perfectly seated and to set the thread pattern. Correct stretch for these bolts is 0.006 inch. Tightening to 65 ft-lb usually gives a stretch of 0.0057–0.0058 inch, with another 2–3 ft-lb of torque to achieve the desired 0.006-inch stretch. Once he achieves this, he actually loosens them all and does it all over again. It's all about discipline.

Finally, with the short-block assembled, Schropp rotates the crank assembly by hand several times—looking, feeling, checking, giving his seasoned eyes and hands one last confirmation that all is well in modular land.

It's a tedious process, but one Schropp feels is worth it.

SOURCES

Clevite Engine Parts
1350 Eisenhower Pl.
Ann Arbor, MI 48108
(734) 975-4777
www.engineparts.com

Livernois Motorsports
2500 S. Gulley
Dearborn Heights, MI 48125
313/561-5500
www.livernoismotorsports.com

Chapter 13
BUILDING A BUDGET 5.0 RACE WINNER

A Potential Race Winner and Performance Street Motor In One Package

Text and Photos by David Vizard

Apart from those components used exclusively for the dyno, such as the electric water pump and Barry Grant fuel log, this 339 hp 346 lb-ft 302 engine can be built for $1,432, including oil.

PART I: THE BUDGET BUILDUP

There are few forms of racing where you can expect to put on a trophy-winning performance with a sub-$1,500 engine. One of them is NASA's American Iron road race series. Here the rules call for a power-to-weight ratio of 9.5 pounds per rear wheel hp. The minimum weight of a car at the end of the race, complete with driver, is 2,800 pounds. This means, for a minimum weight situation, having no more than 295 rwhp.

With a nicely broken in (or nearly worn-out) transmission that equates to some 340 flywheel hp. The question as to whether or not a 340 hp engine in a Mustang can win a race is already answered—it can. This leaves us with only one other question of significance: Can you build, or more precisely, rebuild an engine that makes the requisite amount of power for $1,500? Although a little luck with the basic parts situation (block and crank) is involved, the answer is yes.

In this section I will tell you how, but that won't be the end of the story. Using this basic engine as a starting point, some other very cost-effective options will be investigated. Maybe you don't want to go road racing. Okay, for about $500 more, this cheapo motor's power can be bumped to almost 470 hp and that should make a good showing at any dragstrip. Want something more than a 302? No problem. How about a 331 making over 390 lb-ft and still on production line iron heads—all for as little as $800 more?

Block Selection

To make all this work you need a good block and crank. Fortunately I was lucky here because the old and tired 140K-mile unit from my 5.0 GT had, for some reason, not worn the bores or crank to any significant degree. If you are looking for a motor from the wrecking yard, then be aware that injected units suffer far less bore wear than the older carbureted ones.

After cleaning with Easy Off oven cleaner, good use was made of a wire brush to remove the remnants of all old gasket material.

After masking up and painting, the project engine actually started to look good.

A before and after piston shows how effective oven cleaner can be as a parts cleaner.

The first operation of the cylinder heads was to have the chambers swept out and the valve throats enlarged for the bigger intake valves

One of the last jobs to do before blowing everything off with an air line is to clean out all the block threads. Note how well the bores came out from a simple bore brush operation with a hand drill.

Although little wear was present, the engine was, both inside and out, incredibly cruddy. To save the $30–50 it costs to have a block machine shop cleaned, I invested in two cans of Easy Off oven cleaner and that, plus a reasonable amount of elbow work, brought the block up to scratch.

I then wire-brushed the flat machined surfaces and ran a bore brush through the bores to de-glaze them. Then, using 180-grit emery wrapped around a flat piece of heavy steel, I lightly dressed the decks and other machined surfaces. Next, all threaded holes were re-tapped, and finally the block was given a coat of paint.

The next item for attention was the crank. This needed no work other than putting it through a good session in the parts washer. From here I went on to pistons. Although they had minimal wear, the rings were gummed up to the extent that some were stuck in the grooves. Again oven cleaner saved the day, and, after much effort, the pistons came back almost as new.

At this point the block, crank, rods, and pistons were ready for reassembly. So I now needed rings and bearings. There are two really low-cost sources I can recommend here. First Motor Machine & Supply in Tucson, Arizona, and, secondly, Pro Power in Fort Lauderdale, Florida. Since prices fluctuate, try calling both places for the parts you want.

When ordering piston rings, be sure to specify whether the pistons are the metric 1.5/1.5/4.0mm rings or the 1/16–1/16–3/16ths items. Later forged pistons are usually metric. Once the rings and bearings arrived, the rotating assembly was installed into the block. Attention was now turned to the heads.

Pocket Porting Production Heads

One of the 5.0's weakest points is its cylinder heads. Replacing these with true performance items is outside the budget, but with some effort and a die grinder, it is entirely practical to improve the stock ones. The original heads had 1.78 and 1.46 valves. A set of larger used 1.84 intake valves in good shape from a 305 Chevy were found by going through a scrap parts bin at a local machine shop.

Since no one seemed to want these valves, I got them for the price of a six-pack. The guides in the heads, though worn, were still within tolerance. The first job toward improving the breathing on these heads was to have the machine shop sweep out the chambers around the valves and to cut the throat diameter of the inlet out to suit the larger intake valve. This was the starting point of the pocket porting that was done on these heads.

Essentially all the real porting that was done was to round out the short side turn, slim down the guide bosses, and to blend the bowls of both the intake and exhaust into the rest of the port and tidy up the chambers. Other than removing any obvious lumps and bumps, plus a very basic cleanup, the main section of the ports remained stock in shape

In this shot you can see what was done to the bowl area of both the intake and exhaust. Most of the time spent here is taken reshaping the guide bosses.

Flow numbers were run both before and after the head porting was done.

Seen here is the extent of the chamber work on the heads. The area immediately around the plugs was left as is.

FLOWBENCH NUMBERS

Lift	Stock In. CFM	Mod In. CFM	Stock Swirl	Mod Swirl	Stock Ex.CFM	Mod Ex.CFM
0	0	0	0	0	0	0
25	12.9	15.2	21.7	9.0	6.9	8.0
50	26.0	29.4	32.0	16.4	19.5	21.2
100	54.3	59.0	52.5	55.0	38.2	41.9
150	83.1	89.9	64.0	66.0	55.2	66.8
200	110.5	120.1	69.0	75.0	73.4	85.7
250	130.0	144.8	68.5	77.0	88.4	103.4
300	143.7	168.9	68.0	91.5	101.6	120.0
350	152.2	178.8	68.5	133.0	109.0	134.0
400	157.0	180.0	99.5	146.5	112.1	138.0
450	154.0	182.0	153.5	170.0	113.7	141.0
500	154.0	182.9	167.0	185.0	114.5	142.0
550	155.0	183.0	176.5	200.0	115.3	142.0
600	156.0	183.3	183.0	210.0	115.8	142.0

Although high lift numbers still fall a little short of even small-valve aftermarket aluminum heads, the gains made in the 0 to 500-thousandths lift range that will be used are significant. This porting work probably accounted for some 35 hp on the project engine.

and size. Work on the valves was limited to blending the 45-degree seat into the back face of the valve and giving the valves a reasonable finish. The nearby shot shows the extent of the bowl and chamber work. Figure it will take about a weekend and a half to do the head work at home.

After the principle grinding had been done on the heads, they were returned to the machine shop where the seats were finish-cut, the tops of the guides cut for positive stem seals, and face-milling 40 thousandths off for more compression.

After check-lapping the valves to the seats, the heads were put on the flow bench and tested for me by crew chief Mervyn Bonnett. The numbers were up by a healthy margin, though still just shy of a set of "as-cast" aftermarket heads. Swirl was also good, so some more-than-reasonable results were expected. The graph above shows the bench test results. Remember these heads were ground without the benefit of a flow bench, except to check the before and after numbers. Simply following what was done here should get you very similar results should you decide to go this route.

The heads then went together with a set of single springs and stem seals from Pro Power and were then set aside while the question of the cam and valvetrain was addressed.

Camshaft Selection

We are not working with what could be described as a choice selection of engine components here. Better heads are out of the question because of cost. The same cannot be said of cams. It costs exactly the same to put the wrong spec cam into a motor as it does the right. Cam selection should start with a decision on the Lobe Centerline Angle (LCA), not the duration.

For the record, regardless of what else you may hear and from where, the LCA is not an adjustable feature unless you deliberately want to dial out some power and torque for the sake of a wider power band and idle quality. For any given mechanical spec only one LCA is optimal. If you should hear the comment that a certain LCA is too tight for a street application (it won't idle well and will lack low-end drivability), then be aware that it is not the LCA at fault but the amount of overlap. The fix for this is less duration, not a widening of the LCA.

Having said that, it's also worth pointing out that A) the optimum LCA is mostly dictated by the heads low lift flow capacity and the displacement of the cylinder it is feeding and B) it is better to err on

Do not assume the cam timing will be okay. If you do not know how to time in a cam, be aware that most cam companies have full instructions either in their catalogs or on their websites.

With the cam having four degrees of advance ground in, a stock-style timing chain delivered the required intake centerline setting. Note also the stock FoMoCo oil pump.

a 282 Magnum solid grind from Comp. So long as you stick to the 108-degree LCA, you will get similar results from a 280-degree cam from any reputable cam supplier.

Valve-to-Piston Clearance

The reason a cam no bigger than 282 was chosen was dictated as much by piston-to-valve clearance as it was compression ratio. A 282-degree cam on a 108 LCA with 4 degrees of advance and delivering some 500 thousandths of valve lift should go in okay, but that cannot be taken for granted. To establish this as a fact, the cam was first timed in to a 104-degree intake centerline. The chances were that a stock pattern timing chain set might not deliver the required 104-degree setting. Fortunately with the 4-degrees advance ground into the cam by Comp, a stock-style non-adjustable timing chain delivered a 103.5-degree intake centerline. By the time the chain has broken in, this will be just about right.

At this point clay was placed in the valve cutouts of a piston, and the head and the valvetrain for that cylinder dummied up to establish the piston-to-valve clearance figure. I needed this to be a little more than would normally be the case because I intended to test rockers early on with this combination. As can be seen from the nearby shot, the cam/valve combination, with a stock rocker, not only had enough for immediate needs but also enough for a rocker 0.1 of a ratio higher.

Pushrods and Rockers

Although the block/head combination was 40 thousandths shorter than stock due to head milling, I elected to stay with the stock pushrods as they still delivered an acceptable sweep pattern across the tip of the valve. The rockers for this engine were the best of two high-mileage sets. The pads were dressed the minimum required to smooth out the wear notch caused by the end of the valve.

The rockers bearing surfaces presented a challenge. Normally all of the rockers I used would have been throwaway items. The only reason I could use them is that I was going to lace the oil with Oil Extreme. I've mentioned this stuff before as a recommendation, but here it is near essential—otherwise those worn stock rockers will fail in short order.

Other than wear, the question of valvetrain adjustment needs to be looked at. Here I used

This is the extent of the clearance (about 1/8 inch) that existed after the cam was timed in. The amount available will allow testing of higher lift rockers later in the game.

the tight side than the wide side.

All the foregoing is leading up to the fact that if this engine's output goal is to be met with the hardware being used, then we cannot afford any error in specing out the cam. For a 5.0 with the compression (about 10:1) and flow we have here the optimum LCA is about 108–109. When we stretch this motor to 331 inches, the increased displacement plus the increased compression with the current heads will require a cam of 107–108 LCA. So that there will be near-optimum compatibility with both situations, a cam of 108 LCA was decided on.

Yes, I know that a stock 5.0 comes with a roller (in most cases), but I have gotten to the point where fixing dynamic problems with stock rollers is wearing thin. So that the valvetrain had near-zero dynamic problems, I opted for a flat tappet solid cam. If you follow along on this buildup and you want a quiet valvetrain, consider a flat hydraulic.

Duration for this engines cam was to be in the 280-degree range. Most cams in this range—be they Crane, Comp, Isky, Crower or whatever—deliver about 500-thousandths net valve lift (with a true 1.6:1 rocker), and valve springs delivering 110 pounds on the seat and 280 over the nose will get the job done to 7,000 rpm. Our particular cam was

Here you can see the extent of the wear on the rockers used. This can be a problem with the elevated spring loads and rpm this engine will turn to unless you use a break-in additive such as Oil Extreme by Jet Set Life Technologies.

The pivot blocks of the stock rockers were used to tension the Comp Cams studs from the adjustable rocker conversion kit while the Loctite cured.

With a Barry Grant 650 Speed Demon carb, the two-plane Edelbrock Performer produces about 25 lb-ft more torque at the low end while only dropping about 15 right at the top when compared to a race single-plane manifold.

If you replicate this buildup for use with nitrous, then ARP head bolts are a must as is a 1011 Felpro head gasket.

Comps' rail-style rocker kit to give this setup the adjustability required. When installing the new studs from the kit, be sure to use locking-grade Loctite on the studs, and, equally importantly, preload them as shown nearby during the Loctite's curing phase.

Some important points to note here concern the head gaskets and the bolts to be used. If nitrous is not on the agenda then a regular-quality head gasket and stock bolts will get the job done. If you want to spray a version of this engine, then I recommend the use of Felpro's 1011 head gasket and a set of ARP head bolts. This engine was to be injected so the gasket and bolt upgrade was used.

Induction and Ignition

For a road racer, especially one where road holding rather than power is the number one concern, the stock fuel injection is most certainly not the best bet. Other than the fact it is not a good power producer, we also have to allow it is about 16 pounds heavier than a conventional carb and intake manifold.

This extra weight all resides at the front and at a considerable height above the ground. This makes the center of gravity higher and further forward than would be the case with a carb and manifold. These two items represent the largest hunk of money spent on the engine. The manifold was an almost-new Edelbrock Air Gap Performer RPM obtained at a swap meet, and the carb was a 650-cfm Barry Grant Speed Demon. If you shop around you can find these for under $450 new.

If the budget for a complete engine can be stretched to about $2,000, then a drag race version of this

Here my Custom Performance assistant Jon Wilburn installs a 20-inch collector onto the cool Kook's dyno headers. This combo was good for the low- and mid-speed torque without a significant sacrifice to the top end.

engine—delivering nearly 470 hp and a whopping 545 lb-ft of torque—can be built by nitrous injecting it. For this a basic Nitrous Express plate kit was installed. This was connected up to the resident nitrous hardware in the dyno cell.

For a distributor, I had a unit I pulled out of a carbureted 5.0 a while back and rebuilt it for just such an occasion. This cost, with a new cap and rotor from AutoZone, came to the princely sum of $36. However, I had a similar distributor from Performance Distributors already wired into the dyno ignition system, so this is what you see in the photos. The distributor for this application needs to have 12 degrees of centrifugal advance that is all in by 2,800-to-3,000 crankshaft rpm. With initial timing set at 8–10 degrees, this results in 32–34 degrees total.

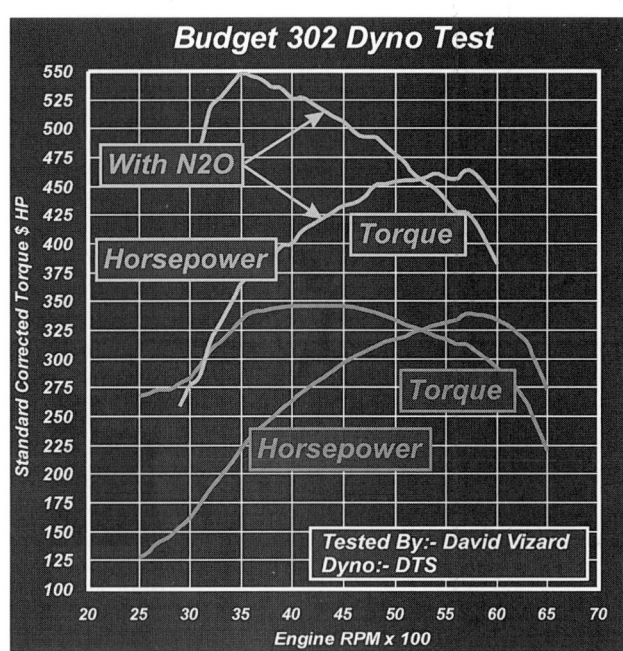

Budget 302 Dyno Test

With N2O

Horsepower

Torque

Torque

Horsepower

Tested By:- David Vizard
Dyno:- DTS

Standard Corrected Torque $ HP

Engine RPM x 100

By any standards this is a very streetable power curve. Within the framework of a NASA American Iron road racer it has the strong torque needed to punch out of the corners. With the nitrous in action, there is more than enough to push a street-weight 5.0 down the strip in the 11s.

With the intake carb and distributor in place, a set of ACCEL plugs (276) were installed along with a set of ACCEL carbon string super budget plug leads. A Fram oil filter and Valvoline loaded with Oil Extreme brought this build up to dyno-ready status.

On the Dyno

Some following this build may not have the option of using long tube exhaust, as per the Kook's dyno headers. If this is the case expect to make about 10–15 lb-ft and horsepower less than you see here. Also those with an eagle eye will have spotted the electric water pump from CSI. I used this as a matter of convenience. Depending on the rpm range, switching to an electric pump appears to deliver between 6–11 hp more. To get the output of this engine with a conventional pump subtract about 8–9 hp.

Other than that there is not much to say here except that the project was a great success. After a two-hour break-in and a change to fresh lubricant, Mervyn Bonnett and I set about fine-tuning the carb and ignition timing. Although usually good with their calibrations, this time the tech guys at Barry Grant really excelled. We pulled two sizes out of the left rear to offset a slightly rich condition that end, and from here every change made reduced power. This corner jet reduction found us about

On the DTS dyno a two-hour break-in period was done to ensure everything was reasonably well-seated. After break-in, the System One oil filter was checked for debris. It proved clear, and the oil was changed to Valvoline 10W40 mineral. This is a good oil for not much money.

2–3 hp so, as an out-of-the-box deal, the carb was very good.

As for output horsepower and torque, peaks were 339 and 346 respectively. With the Performance Distributors vacuum connected to an intake manifold source, idle was what you would expect of a true street unit, as was low-end torque. Throttle response from this combo was more like a sprint car so the cam, head, intake, and carb combo was obviously on the money. Runs on our 2,000hp DTS dyno were started at 2,500 rpm but after setting up, it could be pulled as low as 1,400 with no sign of any stumble that would have resulted had the cam been too big or had too much overlap.

When it came time to shoot the nitrous, it was done just once before finishing off for the day. The NX kit's 150hp jets delivered torque well in excess of 500 lb-ft from 3,500 rpm to 4,600. Peak power increased by some 120 hp, although gains as much as 143 hp were seen lower down the rev range. The drop-off at higher rpm was attributed to the relatively low exhaust flow figures of pocket ported production heads.

With the big torque numbers (545 lb-ft peak), we are in big-block territory and one has to question just how long the stock components in the bottom end might last at this prodigious level. Although this shows that the Nitrous Express system really delivers, this much power on stock pistons and rods might not be such a good idea over an extended period. Whereas these parts may be okay for a season at the dragstrip for a daily driver, I would back off the nitrous to 100hp jetting just to be safe.

The last question is, did all this come in under $1,500? Sure did and by a comfortable margin at that. Even with the machining costs, out-of-pocket expenses only totaled $1,432— excluding the NX nitrous kit. If you figure that into the equation then

Here are the parts that were supplied by Pro Power in Fort Lauderdale, Florida. There was much more to the Scat crank in this kit than just a stroke increase.

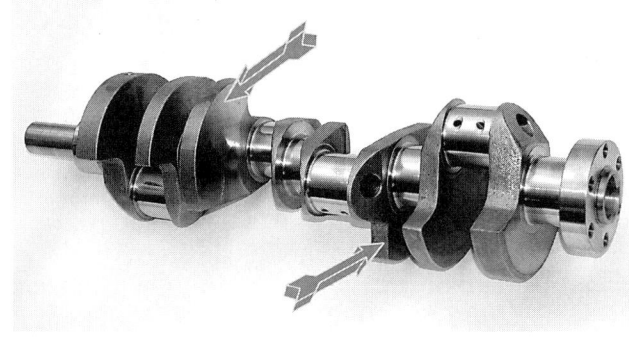

Compared to a stock FoMoCo crank, the Prime One stroker crank has much larger counterweights at the points indicated. The positioning of this extra mass serves to cut the loads on main bearings two, three and four.

The longer 5.4-inch rods supplied by Pro Power maintained the stock rod/stroke ratio even when used with the 3.25-inch stroke Scat crank.

the cost goes up by about $510, but the power is up by a big chunk as well.

Okay where do we go from here? The usual step is to bolt on a set of good aluminum heads, and, while that is still a viable option, let us not forget that we are already pushing the envelop in terms of crank and rod strength with the iron heads this engine is currently equipped with. A set of aluminum heads costs at least a grand. Could this

be spent in some other more constructive way that not only delivers greater output but also addresses the strength of the rotating assembly? If the strength of the bottom end parts is increased, the subsequent addition of a good set of aluminum heads would pay off even better.

What I had in mind here was a stroker kit that will deliver cubes and reliability.

PART II: PRO POWER'S STROKER KIT

The original idea was to build an as-new 339hp, 346 lb-ft 302 for less than $1,500. Mission accomplished. But with an NX nitrous kit, this cheapo motor cranked out some 470 hp and 545 lb-ft—an output that put a serious question mark on the mechanical integrity of the bottom end.

Since the life of this engine was important to me (it was a near perfect 302 block and rotating assembly dyno mule), my first thoughts were to back off the amount of nitrous shot into it to about the 100hp level and have the engine live for as long as I might need it as a test unit. What changed my direction here was a Pro Power ad I saw in our favorite magazine (*MM&FF*) for a 331-stroker kit for just $775. An entire rotating assembly, including rings and bearings, at such a price warranted investigation. My main concern was that there are good inexpensive cranks and there are cranks that may not be solving the reliability problem like I wanted.

To find out what parts were in this kit, I called the boss at Pro Power, Dale Metlika, and asked about the parts content. Here's the rundown: A Prime One (Pro Power's in-house brand) 80-60-06 nodular iron 3 1/4-inch stroke cast crank. Because the graphite in this material is in spherical nodular form instead of flakes, this is far less brittle than cast iron and actually behaves more like cast steel and is often referred to as such. It is both strong and has good anti-wear properties, thus making it an excellent choice for an entry level high-performance crank—something basically capable of handling 550–600 hp from a 5-liter engine.

Rounding out this stroker kit was a set of lightweight, through bolt Prime One 5.4-inch long 5140-forged rods, KB pistons, a Prime One moly ring set and King Bearings.

The fact that these parts had alreay been put through the mill by other racers and were made of materials already proven for the job allayed any fears of unreliable parts. Although these cranks are made entirely off-shore, they are 100 percent inspected in the U.S. before being boxed and shipped. We also ordered a Pro Power crank damper.

The KB hypereutectic piston is lighter than most pistons partly because of the shorter pin used.

The D.S.S. block, CNC machined to its new "Level 20" spec, was one of the smartest blocks to pass our way in a long time.

This operation on the CNC Haas cuts the block for rod clearance when a stroker crank is used.

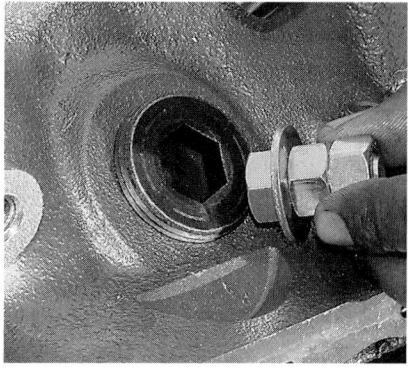

All the drive fit plugs in the block, including the freeze plugs, are tapped to accept screw in plugs.

One crank can be balanced out with only a 28-ounce-inch balance factor. Also this crank has much larger inner counterweights that relieve the inner mains of some of the loads that ultimately lead to block failure. The bottom line here is that we are getting more cubes, a stronger crank, and one that by virtue of its counter weighting also enhances block reliability. The only downside is that this crank is about 5 pounds heavier than stock, but part of this is due to the increased stroke.

The pistons used in our 331 combo were of a cast KB hypereutectic design. The advantages of this type are that the alloy used is hard and resists wear. It can run at close piston-to-wall clearances so it can be quiet. It is slightly less conductive than forged alloys so it holds heat in the combustion chamber, thus it has the potential to produce a little more power. The last point of any major consequence is that a cast piston can be more easily made light as the material can be put exactly where it is needed rather than be dictated by the need to withdraw a forging die from the part after it is forged.

To accommodate the longer rod and stroke within the confines of the 5.0's 8.2-inch deck height, the KB piston has the pin bore intruding into the oil ring groove. Most pistons having this situation to deal with use a steel support rail that is inserted into the groove after the piston has been assembled onto the rod. The KB piston handles this in a slightly different way. Here a pin protruding from the bottom surface of the ring groove locates the lower rail gap so it cannot migrate to the area where the pin bore intrudes into the ring groove.

Design Advantages

The material the Prime One cast-steel cranks are made of is about 50 percent stronger than that of a stock Ford crank. Additionally it has about 50 percent higher fatigue resistance. All this is a good start toward a stronger bottom end, but there's more.

When Ford designed the stock 5.0 crank it considered how to make it using the least amount of material. That's not bad in itself, but it did result in a crank that is deficient of almost any counter weighting for the center pairs of cylinders. For most 5.0s (about 1981 on up) this necessitated the need for external balancing with a 50-ounce-inch out of balance factor. The downside of this is it creates a greater bending moment on the crank snout and the lack of center counterweights exaggerates the main bearing loads at stations two, three and four.

With the Pro Power–supplied damper, the Prime

Block Decision

At this point in the project I came to a crossroads. My 302-mule motor was a near perfect example of a stock 302 bottom end. It is often difficult to get the incentive to build a stock motor and, since this one was in such good shape, I began to think twice about tearing it down to increase its displacement.

A better idea would be to use another block in order to preserve the mule's bottom-end assembly for future testing. A call to Tom Naegele at D.S.S. to ask about a block subsequently proved to be a smart move. Prior to calling I was unaware D.S.S. had recently gotten its upgrade CNC block program online. Designated the "Level 20" block program, this was intended to be the ultimate in

FLOWBENCH NUMBERS

Lift	Stock In. CFM	Mod In. CFM	Stock Swirl	Mod Swirl	Stock Ex.CFM	Mod Ex.CFM
0	0	0	0	0	0	0
25	12.9	15.2	21.7	9.0	6.9	8.0
50	26.0	29.4	32.0	16.4	19.5	21.2
100	54.3	59.0	52.5	55.0	38.2	41.9
150	83.1	89.9	64.0	66.0	55.2	66.8
200	110.5	120.1	69.0	75.0	73.4	85.7
250	130.0	144.8	68.5	77.0	88.4	103.4
300	143.7	168.9	68.0	91.5	101.6	120.0
350	152.2	178.8	68.5	133.0	109.0	134.0
400	157.0	180.0	99.5	146.5	112.1	138.0
450	154.0	182.0	153.5	170.0	113.7	141.0
500	154.0	182.9	167.0	185.0	114.5	142.0
550	155.0	183.0	176.5	200.0	115.3	142.0
600	156.0	183.3	183.0	210.0	115.8	142.0

Although high-lift numbers still fall a little short of even small-valve aftermarket aluminum heads, the gains made in the 0- to 500-thousandths lift range that will be used are significant. This porting work probably accounted for some 35 hp on the project engine.

The arrow indicates the enlarged chamfer area to help reduce the effect of valve shrouding.

would normally expect of a race-prepped block, but also the cleaning and preparation of the block prior to machining produced very nicely finished cast surfaces.

In addition to top-quality machining the D.S.S. block also featured screw-in freeze and oil plugs in all the locations that press-fit plugs are normally seen. Apart from good preparation in terms of machining quality a little extra feature I liked was the enlarged cylinder bore chamfer adjacent to valve. This cuts the effect of shrouding by giving a lead in to the side of the valve nearest the cylinder wall. It's only worth a couple of cfm or so but it's just one more positive feature built into this block. At some $950 a copy this block runs at $300 more than their regular block. That may sound expensive, but for what you get it most certainly is value for money.

Before moving on to the next phase of our build, I should point out that I could have used the mule's original 302 block, but chose not to so the price of the D.S.S. block is not factored into the overall cost if you are looking to replicate what was done here.

Bottom-End Assembly

Once all the rotating assembly parts were acquired they were balanced at More Performance in Charlotte, North Carolina. This company normally specializes in late-model, fuel-injected GM engines, but they were good enough to jump right on this job at a moment's notice. Everything was brought into balance at a 28-ounce-inch external balance factor and a few hours later I walked out of More Performance with a rotating assembly ready to go.

After balancing, the first move was to install the pistons onto the rods then the rings, which had been previously gapped as per KB instructions,

Here are the major bottom end components used in our project motor. Balancing and ring gapping have yet to be done.

automated machine block preparation. Knowing how fussy I was to have things done right Tom was more than ready to have me check out a block and tell *MM&FF* readers what a high-quality piece it was.

It so happens I am very much a "show me" guy and no amount of claims over the phone is going to convince me. The only way to do that is for me to have the hardware right in front of me with all my micrometers and bore gauges at hand.

I am not going to bury you with dimensional details, but this block checked out to be exactly what was claimed of it. If I said it looked as good as new it would be almost an insult to D.S.S.'s work. No new block ever looked this good.

Not only was the machining done within what we

VALVETRAIN TECH

Although we made them work here, I was not really happy about the stock rockers so I decided to do some valvetrain tests as a separate issue. The intent here is to show how an increase in displacement also calls for an increase in valve lift if the best results of the extra cubes are to be realized.

For rockers I chose Comp Cam's budget-priced Magnum rockers. These are roller tipped, but utilize a ball pivot. The particular type used was subject to our Spintron valvetrain. On the Spintron they performed very well and, with the eared tip design, needed no pushrod guide plate.

After lash, the stock rocker, at supposedly 1.6:1 ratio, delivered 475-thousandths lift at the valve. A trial fit with the Magnum rocker revealed the sweep was a little too close to the far edge of the valve. To offset this, Comp's Magnum pushrod, 100-thousandths shorter, was used. The shorter pushrod resulted in nearer optimal geometry and allowed the rockers to deliver an additional 35-thousandths lift that maxed out at .510.

Back on the dyno, the Magnum pushrod/rocker combination proved effective at delivering a substantial amount of extra top-end power. The price paid for this was a small reduction in torque up to 4,100 rpm. After this the torque increased progressively all the way to the self-imposed 6,500-rpm red line.

At 6,500 the additional torque amounted to about 40 lb-ft. Peak power moved up from 5,800 to 6,100–6,200 rpm. Although peak torque only moved up by about 2–3 lb-ft the peak power rose just shy of 10 hp. The results we see here are typical of an engine that has a cam with enough overlap area for the cam duration involved. In other words it has an LCA that delivers the right overlap triangle area. If this is increased it will have too much overlap area and output in the lower half of the rpm range will suffer.

Adding extra lift via the rockers also meant the valve was further open at all points in the lift curve. This resulted in more overlap area to the extent it was slightly too much for our 302. A note here is that the higher the rocker ratio used the wider the LCA needs to be. Conversely a higher rocker ratio will help compensate for a cam with an insufficiently tight LCA.

When the same test was done on the 331 variant of our engine the results were all positive. At 331 inches our engine needed to have an LCA a little tighter than at its original 302 displacement. This would produce an increase in the overlap triangle area to commensurate with the 331's needs.

Rather than swapping out the cam, the alternative was to install higher ratio rockers as this will do the same thing. If the overlap triangle is too small (be it via LCA or too low a rocker ratio) for our 331's requirement then the engine should respond to the higher ratio rockers by delivering a greater output over almost the entire rpm range. There was no loss in torque even down as low as 1,400 rpm. From 3,500 rpm the torque with the higher lift rockers really started to pull away from the stock ratio rocker setup.

Peak torque climbed from just over 390 to just over 400 lb-ft. At 6,500, where this engine is really air starved, the added lift delivered 53 lb-ft more torque. That, at 6,500 rpm, equates to a whopping 65 hp! Peak power with the Magnum pushrod/rocker combo was up from 376 to 393.

So what have we learned from this excursion into valvetrain tech? These tests should have driven home the point that when going to a larger displacement, a cam that may (emphasize the may) have been optimal for the original displacement, won't be for the bigger displacement. A larger engine needs to have a tighter LCA and/or more lift.

onto the pistons. Details concerning the lower oil control rail are vitally important so read the KB instructions if you intend to use these pistons. Also follow the procedure outlined here when installing rod/piston assemblies into the block.

Although all the bottom-end parts had been checked with micrometers, etc., nothing was taken for granted. After installing the ARP extended mains studs (the extensions retain the D.S.S. main girdle) and King bearings, the crank was laid into the block and the clearances given a final check using plastigauge. These indicated main bearing clearances a couple of tenths tighter than we would have liked for a race engine, but were fine for a street motor where extended life and quiet

operation are a factor.

With the crank in and the caps torqued into place, it was time to install the piston/rod assemblies into the block. Here there is great potential to foul up what would otherwise be a super bottom end. Take note of what is said here and all will be well.

First use a tapered ring compressor (Total Seal is a source for these) for this installation, not one of those otherwise functional sheetmetal compressors. Then, making sure you are putting the assembly in the right way and in the right hole, centralize the oil ring while making sure the locating pin on the lower oil ring groove face is centralized in the oil ring gap.

An above-average amount of metal removal was required to achieve the 28-ounce-inch balance factor.

If you don't own a $500 bore mike, then Plastigauge runs a real close second best for checking bearing clearances.

After the rods and pistons are installed the torque settings of all the mains and rod bolts was reaffirmed.

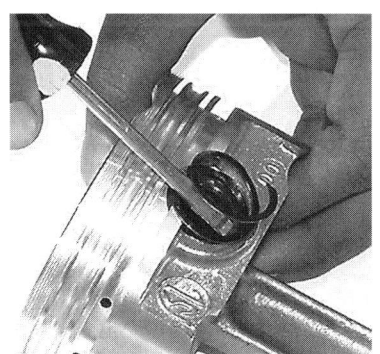

If you experience difficulty installing SpiroLocs then it could be you are not expanding the coils enough prior to installing.

With a pinned lower oil rail it is best to use a tapered ring installer such as this Total Seal item. Regular ring compressors tend to have a problem with ring hang up.

As with the 302, the CC 282 Magnum solid flat tappet cam was timed in at 104 degrees intake centerline.

Now slowly push the assembly into the ring compressor making sure as far as possible that the oil rail has not moved around. If all this slides in and needs no more than gentle tapping from a rubber mallet all is well. If you have any inkling that it is hanging up, remove the piston and start the procedure again. All this sounds fussy, but once you have done it a few times it is simple enough. When complete, the rotating assembly turned over at just 18 lb-ft (a stock new 302 assembly is 28–33 lb-ft).

Parts Swapping and Dyno Flogging

The next move was to install the cam and lifters from the 302 short-block. Before installing the lifters the face was refinished by making a figure-eight motion on 180-grit emery on a surface plate. This restores the correct finish/radius on the lifter so returning it to as-new. The cam installation procedure was done as per the 302 to establish the same event timing. From here it was just a question of removing the heads, induction and ignition

systems from the 302 and installing them on the 331.

After breaking in the new engine for about two hours it was given its first conservative pulls. This involved just short pulls to ascertain that all was well and to determine what, if any recalibrations of the ignition and carburetion were needed. After about three hours of testing, we arrived at the conclusion that this combination needed about a degree to a degree and a half less total timing, along with one jet size larger than had been used in the 302 configuration. As for the results they were more than satisfying as can be seen from the graph on the next page.

CONCLUSION

The success of this project is greater than may initially seem. The displacement went up by 9.6 percent, but the output rose at the low end by as much as 14 percent. Only above about 6,000 rpm did the percentage gain in output drop below the percentage gain in displacement.

So where did the extra output come from? Digging into this there are a number of factors that helped boost the output over and above what might normally be expected. First is that the compression ratio with the KB pistons and the stroker crank was 10.35:1 as apposed to the 302's 10.0:1 so a little extra came from this quarter.

Also helping output was the fact that the piston

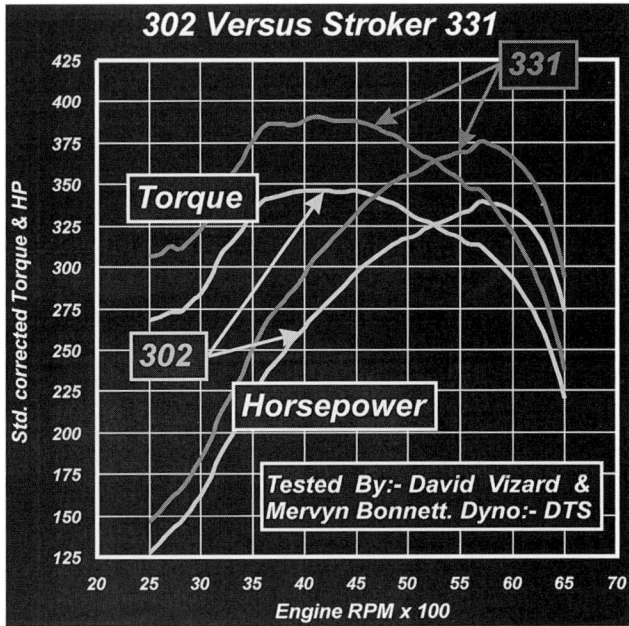

302 Versus Stroker 331

Torque

Horsepower

331

302

Tested By:- David Vizard & Mervyn Bonnett. Dyno:- DTS

By avoiding camshaft pitfalls and selecting parts that tend to aid power our 331 came in with some very rewarding results. Torque was up by 44 lb-ft and horsepower up by 37.

deck height clearance of the 302 was 12 thousandths whereas the 331 was 8 thousandths. Experience has shown that this seemingly small difference can be worth about 3 hp. The hypereutectic pistons have less friction than the OE forgings and they keep the heat in the chamber slightly better. This also helps output in a positive manner.

The block was also very likely a contributor toward aiding power. Precision machining and the bore chamfer being possible contributors here. Lastly there was the cam. Because this was spec'ed to give best results for an engine somewhere midway between the 302 and the 331 we did not fall foul of the problem of having a cam way off the mark for the bigger engine.

With 376 hp and 390 lb-ft of torque for a little over $2,400 (including the crank damper) this looks like one heck of a deal. Not only have we got extra performance, but also a bunch of extra reliability. Now if we were to put on a set of aluminum heads…

SOURCES

ACCEL
10601 Memphis Ave. #12
Cleveland, OH 44144
216/688-8300
www.mrgasket.com

Automotive Racing Products (ARP)
531 Spectrum Cir.
Oxnard CA 93030
805/278-7223
www.arp-bolts.com

Barry Grant Inc
1450 McDonald Rd.
Dahlonega, GA 30533
706/864-8544
www.barrygrant.com

Comp Cams
3406 Democrat Rd.
Memphis, TN 38118
901/795-2400
www.compcams.com

Custom Performance
5255 Pit Rd. South
Concord, NC 28027
704/454-5220
www.gofastfords.com

CSI Performance
16936 C.R. 252
McAlpine, FL 32062
800/226-1247
www.csiperformance.com

D.S.S. Racing
3550 Stern Ave.
St. Charles, IL 60174
630/587-1169
www.dssracing.com

Edelbrock Corp.
2700 California St.
P.O. Box 3500
Torrance, CA 90503-3907
310/781-2222
www.edelbrock.com

KB Pistons
United Engine & Machine
4090 Goni Rd.,
Carson City, NV 89701
800/560-4814
www.kbpistons.com

More Performance Inc.
2919 Interstate St.
Charlotte, NC 28208
704/393-8241
www.moreperformanceinc.com

Motor Machine & Supply
1401 W. Glen
Tucson, AZ 85705
520/792-1156

Nitrous Express Inc.
923 Lake Park Dr.
Wichita Falls, TX 76302
888/463-2781
www.nitrousexpress.com

Performance Distributors
2699 Barris Dr.
Memphis, TN 38132
901/396-5782
www.performancedistributors.com

Propower Performance Parts
4750 N. Dixie Hwy., Ste. 9
Fort Lauderdale, FL 33334
954/491-6988
www.propowerparts.com

SCAT Enterprises Inc.
1400 Kingsdale Ave.
Redondo Beach, CA 90278
310/370-5501
www.scatcrankshafts.com

Total Seal Inc.
11202 N. 24th Ave., Ste. 101
Phoenix, AZ 85029
602/678-4977
www.totalseal.com

Chapter 14
COST CONSCIOUS CUBES

What Happens When You Take a Set of High Flow, Big-Port, Big-Engine Heads and Put Them On a Smaller Engine?

Text and Photos by David Vizard

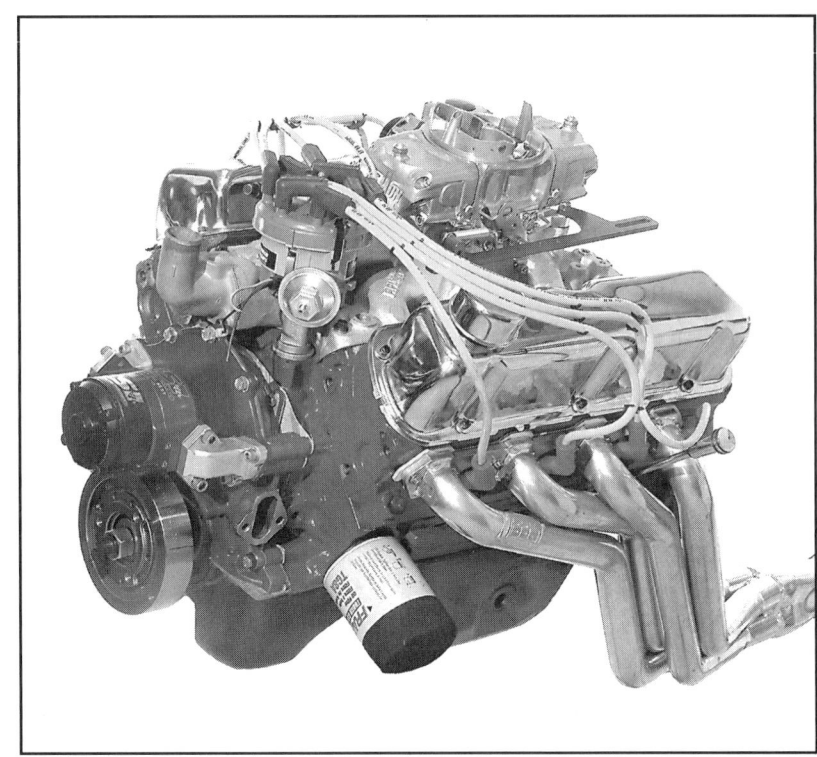

This UNCC student-built, totally stock long-block engine cranked out a respectable 317 lb-ft and 287 hp with a Victor Jr. intake and a Barry Grant Road Demon carb.

PART I

Most of those who contemplate building a large-displacement 5.0 are unlikely to use a set of stock heads on it. The engine in this story is destined to use a set of Dart Pro 1 heads with 195cc ports and big valves more suited to a 347-inch bottom end than a 302. Here's the question being posed: What if the heads were acquired before the bigger bottom end was built? Could they be used effectively on the original 5.0 until such time as they were needed for the new bottom end?

In the interests of generating a more realistic baseline to find out what effectively utilized cubes are worth, this is just what we did.

Engine No. 1

Since our 347 was intended to run an Edelbrock Air Gap Performer RPM or a Victor Jr. intake, along with an appropriate carb, the first priority was to establish what an otherwise stock 302 long-block would do with both of these performance-proven intakes. For this, a long-block was built by a small group of UNC Charlotte Motorsports students. The base engine was a 160,000-mile unit that had only minimal wear. The worst bore was 0.0003 inch worn, and all the bores honed up clean to the desired size for 0.003 inch piston/bore clearance.

The piston ring grooves were like new, and the piston skirts had worn only about a tenth of a thousandth or so.

The crank was indistinguishable from new. With the block rehoned and the crank polished, the students went ahead and performed a meticulous build. Basically they went about it with the same diligence to fitment, limit, and clearances that you would expect to be applied to a full-fledged $45,000 Cup engine. This point is being made because the quality of the students' long-block build was probably worth

Stock ring widths, but in Total Seal design, were used in this build. They are a simple power adder and have a long life, but due care must be paid to installation.

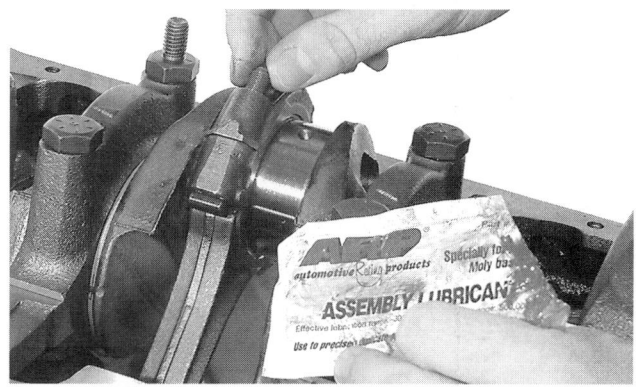

The rods used in our baseline engine were reconditioned and equipped with ARP bolts. As per ARP instructions, the bolt and nuts received a healthy dose of ARP thread lube before being torqued up.

The stock replacement Professional Products damper was installed as any damper should be—with a proper damper-installing tool. For the record, a hammer is not an approved tool for this job.

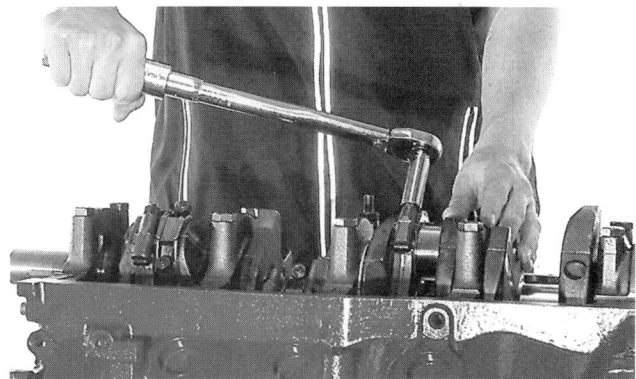

To establish that there was good alignment of the main bearing, the crank was first installed, using light oil, without the rear oil seal. From here, each cap was tightened, and the turning torque checked. Any bearing that showed up stiff would be grounds for investigation.

Initial tests were done on KT Engines' dyno in Concord, North Carolina. Thanks go to boss Kenny Troutman for allowing us to use the dyno on such short notice.

10 hp over one fresh from the factory.

To give you an idea of how well they succeeded, the short-block turning torque, with stock tension Total Seal rings, was only 18 lb-ft. That's a creditably low number compared to a typical factory fresh build at 28–30 lb-ft.

Just for the record, the bearing clearances were, by a process of bearing selection, set at 0.0022–0.0024 inch for the mains and 0.0018–0.0020 inch for the rods. Even the lifters were stripped and rebuilt. Any that collapsed too quickly would be rejected. In addition to this, cam timing was checked to see if it was where it should be for best output. That, incidentally, is usually a couple of degrees more advanced than as it comes from the factory. Because the stock damper was so old and pretty much used up, we elected to replace it with a new stock replacement from Professional Products. This looked well-made and the price was affordable.

For the induction system, an Edelbrock Air Gap

Performer RPM was installed for the first round of testing. On this, a BG Road Demon was mounted along with a smart throttle linkage bracket from AED.

At this point, we were ready to introduce our baseline engine to the dyno. Previous dyno commitments prevented us from using the UNC Charlotte Superflow dyno, but Kenny Troutman of KT Engines in Concord, North Carolina, saved the situation. By allowing us to come in over the weekend, we managed two full days of testing without upsetting his normally busy schedule too much.

The results achieved from this build showed that care taken to do things right during assembly does pay off. Just check out the curves on the dyno sheet. Along with its good WOT performance, this carbureted 5.0 produced a smooth 575 rpm idle with, on the Performer, a manifold vacuum of over 17 inches. This says much for the high functionality of the carb and intakes used.

What we have done so far is establish two baselines that we can make a comparison with later in terms of manifolds and what the worth of a bigger inch motor might be when built with

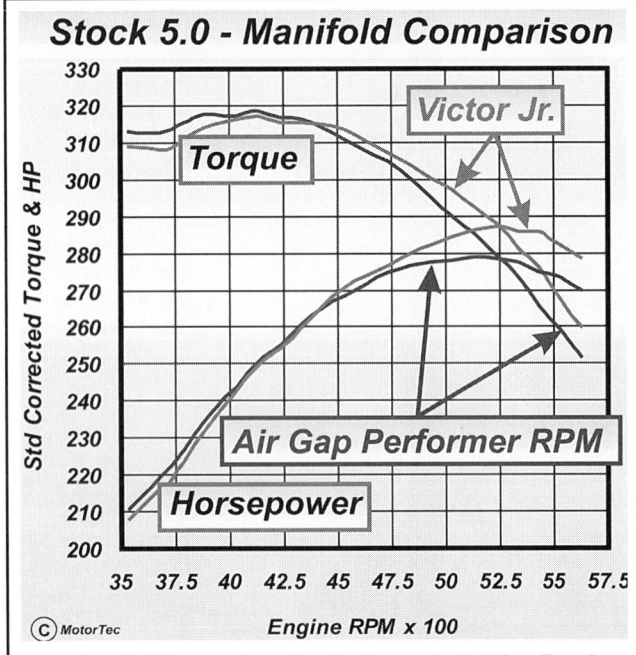

Stock 5.0 - Manifold Comparison

With an all stock long-block, our baseline engine produced the curves seen here. Highest torque, produced with the highly functional Air Gap Performer RPM, was 318.7 lb-ft. The Victor Jr. lagged this only marginally at 317.1 lb-ft. The situation for top-end power was, as would be expected, reversed with the Victor Jr. showing 287 hp against the Performer's 279 hp. For the record, a good long-block like this tops out at about the 316 lb-ft and 255hp mark with the 5.0's stock fuel-injected manifold. Who says carbs are outdated?

The otherwise stock heads on our second engine were converted to accept aftermarket studs to utilize Crane's 1.6:1 Gold Race Rockers. For this round of tests, the Edelbrock Air Gap RPM Performer intake was used.

appropriate hardware. We are now at the point where we can ask whether it is practical to use a set of heads that are basically intended to be used on a bigger-inch engine than we currently have. This is what we will now investigate.

Engine No. 2

The second baseline engine was one from the author's stock that had previously been used to test street-type cylinder heads and manifolds. It was equipped with a typical street-style cam. This was a Comp Cams 3108 hydraulic roller grind from its Magnum series. It spec'd out to 271 degrees of off-the-seat duration and 215 degrees at 0.050 inch. The lobes were ground with a 110-degree spread and at 4 degrees of advance. Lift with 1.6 Crane rockers was 0.533 inch. This profile was chosen because it is easy on the valvetrain yet generates enough lift to take advantage of a big-valve cylinder heads' high-lift breathing advantage. (For the Dart Pro 1 flow test results, check out the sidebar on page 105.)

As you can see, the flow figures climb quite steeply until about 0.600 inch lift. This particular head has 2.02 intake valves, which will not clear the valve notches without some additional machining. Many installation shops have a dummy head and a cutting tool that allows them to enlarge the valve pockets by the relatively small amount needed to clear bigger valves without removing the pistons

from the short-block. In our particular case, the short-block was equipped with Ross pistons with the appropriate valve clearance for the job.

At this time, we were ready to dyno test. For this and all our subsequent big-inch tests, we used the T&L facility in Stanfield, North Carolina, for a number of reasons. First Lloyd McCleary, the boss at T&L, in readiness for this big-inch bash, set aside one of his three Superflow dynos for our project. It was not quite exclusive for us to use, but for most practical purposes it was darn close to being so. Secondly T&L does a lot of 347-inch engines from mild to wild and from bargain basement prices to Busch car prices. Couple this with the fact that McCleary builds all his engines as if they are going out to race. This means that all his engines, whether they are street cruisers or race units, go on one of T&L's dynos to establish they are doing what they are supposed to before being shipped.

After a good break-in on our Air Gap Performer RPM–equipped mule motor, the oil and filter were changed, and testing started in earnest. Because the stock-headed setup was a known combination, dialing it in took only a matter of minutes.

After a cooldown, the stock heads were removed. New FelPro gaskets were then installed, and the Dart heads were dropped into place. The ARP head bolts were given the sealer treatment and torqued down. Lastly the intake and carb were replaced. Within less than an hour, we were ready to run our Dart-equipped mule.

Although Lloyd had run heads of around 170 cc on 302s, he had never run any as large as 195 cc on 302s, other than high-winding race-spec engines.

DART'S PRO 1 HEADS

Dart's Pro 1 heads come in two versions, the big valve/port version as seen here, and a smaller valve/port version intended to be a bolt-on for a 302 with stock pistons. The big-port ones have a 2.02 intake fed by a 195cc runner on the intake and a 1.6 exhaust valve dumping into a 75cc exhaust port. The smaller version has 1.94 intake valves and 170cc intake ports, but shares the exhaust valve size of the bigger head (though this dumps into smaller 65cc exhaust ports).

The exhaust ports of both heads are raised by 0.135 inch and utilize a standard Ford bolt pattern. As for valve lengths, there is a difference between those required to be used with a solid roller compared to those for a flat tappet or hydraulic roller. The valves for the solid roller are plus 0.100 inch to accommodate the heavier spring required for a roller cam. Retainers for whatever springs required are the Comp Cams 10-degree type with machined locks. If you're bolting these heads onto a 5.0 block (as opposed to a 351 block) the ARP stepped-head washers/bolt hole reducers (PN 154-3605) will be needed.

Moving on to the all-important flow bench tests, we find the 195cc Dart Pro 1s are strong performers. Not only did the ports flow well at high lift considering they were "as-cast," but they also delivered better than average low- and mid-range flow. These numbers are important especially when the valves are being lifted less than to the point where flow starts to top out. In other words, airflow that occurs outside the accessible range of the valve lift used serves no good purpose.

For the more adventurous home engine builder, another good aspect of the Dart Pro 1 is the fact that they are a piece of cake to do a professional-looking port job on. All you need do is streamline the valve guides to the small amount possible and clean up the shape that Dart built into them. If you spend a couple of weekends doing just this, the result is about a 12-cfm increase at 0.600 lift with gains starting at about 0.300 lift.

As to how much power this may be worth, it's hard to say, as it depends on how air-starved the combination is that it runs on. But there is more to it than just airflow. The fact that porting these heads only increases the port volume by about 3 cc shows how little needs to be removed. The big deal here is that the smoother ports have less surface area than the cast-finish ports, and, as such, less heat is conducted into the intake charge. This alone is worth about 8–10 hp on almost any application.

Dart's Pro 1 heads can be had as complete assemblies sprung ready to install or as partial assemblies, such as the finished casting complete with valves ready for your custom valvetrain.

The Dart combustion chamber form leaves both the intake and exhaust valves with just about zero valve shrouding. This means the only valve shrouding is that unavoidably caused by the cylinder bore wall.

The importance of the port and chamber shape in the vicinity of the valve seats is critical to good flow. Here we see that the intake is blended to the seat and that the form immediately under the exhaust seat mimics that used in many Pro Stock and cup car applications.

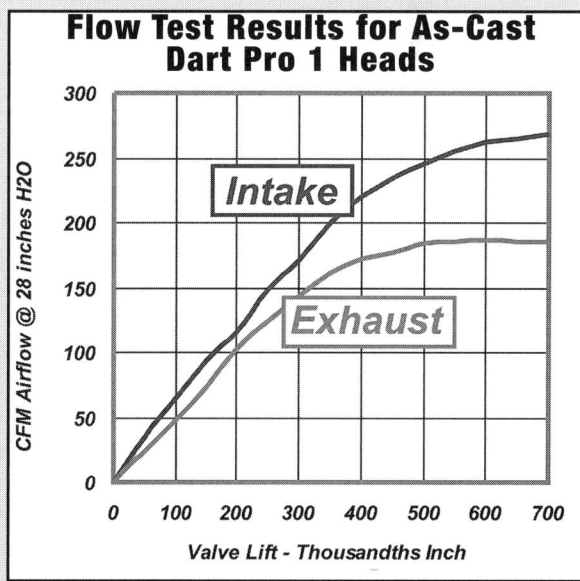

Flow Test Results for As-Cast Dart Pro 1 Heads

The Dart Pro 1 flow figures are strong for an as-cast head. The intake flow fell shy of the 270-cfm mark by a narrow margin. The exhaust topped out at almost 190 cfm. To put that into prospective, there are some head castings on the market that, when given a good but basic port job, deliver numbers no better than this.

Our No. 2 baseline engine started off life as you see here. The only difference on the dyno was a Moroso pan and a Pertronix distributor.

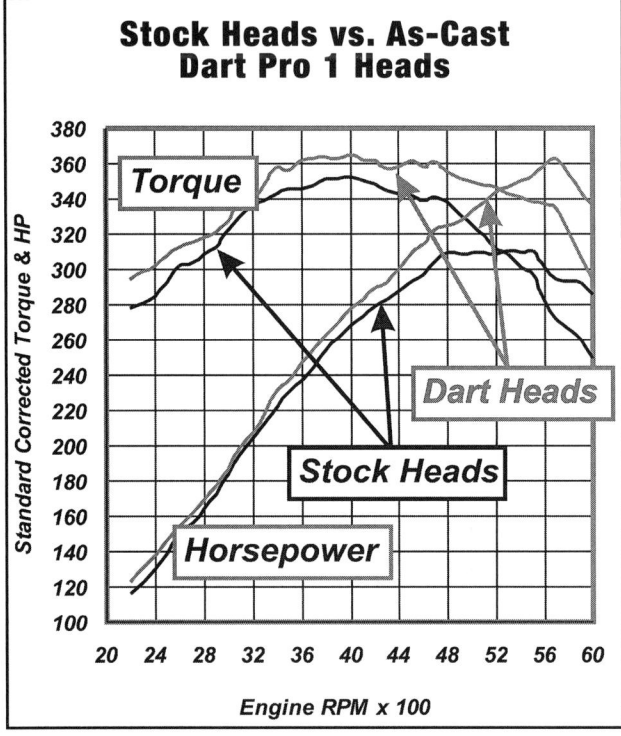

Stock Heads vs. As-Cast Dart Pro 1 Heads

Torque

Dart Heads

Stock Heads

Horsepower

Standard Corrected Torque & HP

Engine RPM x 100

The Dart 195cc port Pro 1 heads in combination with an Edelbrock Air Gap Performer RPM produced excellent low-speed results along with a 52hp increase at the top end.

(For the record, these 195 Dart heads actually measured out nearer 190 cc, but we will continue to refer to them as 195s to stay in line with Dart's listing of them.)

In spite of this, the required jetting and timing were quickly achieved, and some impressive results were produced. But there is more to the story than just the raw numbers. First, we need to remember that a good engine combo is what gets results, and although we knew going in that the 195cc Pro 1s were a little on the big side, it appears that using a good two-plane manifold as we did here helped offset the effect of the bigger-than-desirable port.

Nevertheless, the fact that the Pro 1s produced more torque everywhere in the rev range proved that in the case shown here, it is possible to utilize heads intended for a bigger engine on a smaller one. These were good results, however, based on not

only his experience with numerous types of 5.0 aftermarket heads but also extensive experience with Dart heads in particular, I asked Lloyd for his take on the results. He ventured that had we used 170cc runner volume heads like the smaller Dart Pro 1s, we would have most likely seen about a 5–7 lb-ft increase along with about an 8-or-so-horsepower increase. But if such heads were used on anything but the mildest cammed 347, we would see a reverse in the trend from the mid-rpm range up compared with the 195cc heads.

At the end of the day, the fact is the big-port Dart heads were so much better than stock that they produced positive results across the board even though they were not optimal for a 302. If you intend to stick with a 302 bottom-end, maybe you should put the 170cc Dart heads on your list to seriously consider.

As for the 195cc Darts, we're looking forward to using them on our upcoming 347-inch build, and running some cam and intake comparisons. Our target for this will be to build the lowest-cost 347 without sacrificing the use of proven parts.

PART II

If you reread Part 1, you will remember the premise was whether or not it was feasible to temporarily install heads intended for a larger-displacement engine onto your current 5.0 and still get positive results. We did just that with our 195cc Dart "as cast" heads and picked up over 50 hp as well as pushing up peak torque by 10 lb-ft.

In this section, we are essentially going to take that same 306-inch engine and make it into a 347, with the intent of establishing more precisely what can be gained by stretching the engine an extra 41 inches.

To upgrade a 302 to a 347 at the time of an overhaul minimizes whatever additional costs are involved. All that is needed are different pistons, the stroker crank and rods, and some additional block machining. Acquiring a Scat cast steel crank, forged race spec rods, and a set of lightweight Ross pistons from T&L Engine Development will set you back an additional $635 at most over and above a regular rebuild cost. (T&L has a whole rotating kit balanced and complete with bearings and rings for $899.) If your current 302 crank needed a regrind, then the cost difference drops about 100 bucks.

In addition to getting extra cubes, the new crank, rods, and pistons are far stronger parts, thus allowing more rpm in spite of the longer stroke. Still, our first move was just adding the cubes—13.4 percent of extra inches to be precise. Our aim was to find out what we can expect to get back in terms of percentage of additional output.

T&L puts in as much effort on the bores of a regular mail-order street motor as it does on a Cup car motor. This means you're buying top quality.

The ARP bolted Scat rods are tough, affordable, and, among the lower-cost rods, up to 30 grams lighter than some you might see.

The flat-top Ross piston design used weighed in, with pin, at 517 grams. That is as much as 90 grams lighter than some pistons sold for performance applications.

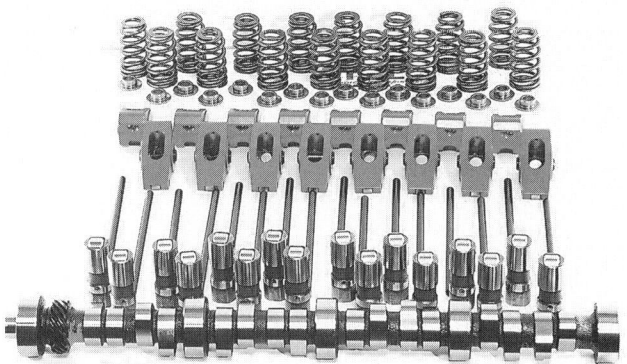

Here is the Comp Cams valvetrain used for our tests. The cam was 276 x 282 seat duration with 0.513 valve lift achieved from a 1.6:1 rocker ratio.

Scratch Built

For doing tests such as this, it is easier to start a 347 build from scratch by using an identically prepped block and installing the new stroker parts into it. Doing just that, our new 347 rapidly took shape in T&L's shop. The bores got the standard T&L race-quality prep deck plate hone on their Sunnen hone. Along with this, the mains were align-honed and the block decked so the Ross flat-top pistons would be flush to the top of the block. This beautifully prepped block then received several coats of primer and black engine enamel.

For pistons, we used a Ross off-the-shelf design. Ross has an industry reputation of producing lighter-than-average pistons, and ours, PN 99822 at +0.030 over, was no exception. Each piston itself weighs in at 399 grams, but a further weight savings, compared to more production-orientated pistons, comes about from the use of a shorter-than-normal wristpin. The norm for production pistons is to use a 2.850- to 3.00-inch-long wristpin. By minimizing the spread between the pin bosses, the Ross piston can effectively use a 2.750-inch-long pin without giving away any effective bearing area. The result is a pin that is 8 percent lighter right off the bat. Add to this the fact the shorter pin is stiffer so it needs less wall thickness, and we see pin weights that drop from the 140–150 gram mark down to the 118–120 mark.

Having the correct cam timing is important and becomes more so the more optimally spec'd the cam is.

Head gaskets can be a problem with high-compression ratios in a Windsor-style engine. We use Fel-Pro as a good insurance against failures.

Here, our Dart Pro 1 heads are being positioned on the new block.

Because of the smaller head-bolt threads on a 302 compared to a 351, stepped ARP head-bolt washers are required (arrow).

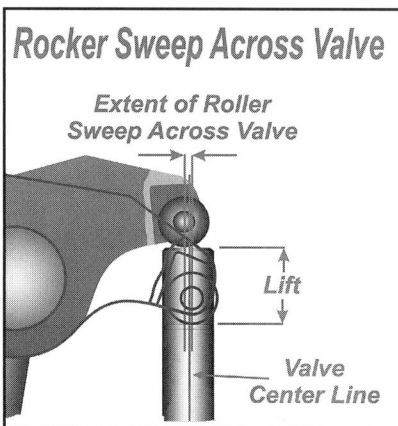

Rocker Sweep Across Valve

Extent of Roller Sweep Across Valve

Lift

Valve Center Line

Pushrod lengths need to be selected such that the roller sweeps out a patch centralized on the valve-stem end.

If you want to spring for special tool steel pins, then pin gram weight can drop to the mid-90s. This may sound like only a few grams, but it can make the difference between having to add heavy metal into the crank to achieve balance or not. In practice, a few grams here can make a $100 difference in the cost of the balance job. At the end of the day, cost and quality make the Ross piston a good choice, and performance on the job comes as a bonus.

Rods are a delicate subject, especially if you are a manufacturer that cares about reputation. It's one of those critical engine components that if one breaks, it usually takes out pretty much the entire engine. This scenario gets more likely the better you are at building power. When you start dropping down the price scale, know what you are buying. Within the realms of the cost-conscious rods, there are purchasing pitfalls. Sure, a lot of different brands all have the same look, but does looking at them give you any idea about the quality of the material used to make them? They can say 4340 on the box or in the literature, but that is no guarantee they are up

to scratch in that department.

We have used Scat's low-cost rods of various patterns since they started making them some 15 years ago. Scat has been in the crank business at both ends of the price scale forever and a day. In our case, that's a race-spec H-beam rod from Scat where they sonic test, X-ray, and Magnaflux as necessary to ensure the best reliability.

Valvetrain

The valvetrain used was virtually a replica of that used in our 302 baseline motor. This was a Comp Xtreme Energy 276 by 282 profile, and for the record, this cam, which delivered 0.513 inch lift with a 1.6:1 rocker, was optimized, in the 302. From here, the valve motion went through the same Comp hydraulic rollers, pushrods, and rockers as did our 302.

At this point, the piston/rod assembly, less rings, was temporarily installed into the No. 1 cylinder just for the purpose of timing in the cam. With the intake centerline at 106 degrees, the cam delivered the same valve events as were found optimal in the 302.

With the cam timed in, the rings, which had previously been gapped, were installed on the pistons, and these, in turn, installed into the block. From here, our Fel-Pro head gaskets were positioned on the block and the same as-cast Dart Pro 1 heads as used before were bolted on using ARP hardware.

The pushrod length was checked to see that it produced the correct rocker sweep across the valve tip as per the illustration at left. After establishing, with an adjustable pushrod, the length that

Here is the valvetrain with the correct pushrod lengths for best antiwear properties as well as best power production.

The fit and location of each rocker in relation to the valve is checked as the valvetrain assembly progresses.

The securing bolts of the Edelbrock Air Gap Performer RPM intake manifold are progressively snugged up before a torque wrench is used for the final series of tightening sequences.

A neat plug wiring job not only enhances the motor's appearance, but also is a chance to route them as far as need be from hot exhaust-system components.

delivered a centralized rocker contact patch, pushrods of that length were installed. This is another one of those time-consuming ops that an engine builder aiming for top quality will do, but is not necessarily done with the average crate motor build. Next, the rockers were assembled onto the ARP rocker studs and the lash adjusted a quarter turn into the lifters. With that, the valvetrain was done.

Induction and Ignition

T&L offers a variety of carb and manifold options for its 347s. All have been included in the lineup only after they have been dyno tested and shown to have desirable properties. In terms of intake manifolds, T&L's Lloyd McCleary tells us there are a number of good Holley and Edelbrock intakes for a carbureted 347. In this instance, we are using the very much proven Edelbrock Air Gap Performer RPM as per our baseline engine. For a carb, we ran the dyno room Holley (approximately 800 cfm) as per Round 1 of our big-inch build. What you may want to use for a 347, if you were buying one, depends on the application. If it were a street cruiser where the absolute biggest top-end numbers are not of any real importance, a shiny 600-cfm Edelbrock would be a good choice. If street and strip performance are issues, then, depending on the intake manifold used, a Holley or Demon carb in the 750- to 850-cfm range would be more in line with what is needed.

For sparks, the same PerTronix distributor as per our baseline engine was used to fire the dyno's MSD ignition system. The coil output was transferred to the Autolite plugs via ACCEL plug cables. And, in passing here's something of note. Although Autolite plugs are near the bottom of the price scale, the dyno shows they run near the top of

the performance scale. If those $12-a-pop plugs are out of your price range, consider Autolites. At less than $2 each, they're a viable alternative.

Dyno Time

A two-hour break-in was done before any serious pulls were made on our 347. Since McCleary had built and dyno'd dozens of these engines, the carb jetting was pretty much a known quantity, and, if anything, only some fine-tuning was expected to dial it right in.

The same applied to ignition timing. As a result, the first pull gave us a clear expectation of final results. After a couple of foreshortened pulls to establish all was well in terms of zero oil and water leaks, our 347 was given the go. After three pulls, it was apparent this was a highly repeatable engine as each pull delivered curves close to the others. The curves on page 110 clearly show the gains made with our stroker 347.

By stretching the motor from 306 inches to 347, we increased the displacement by 13.4 percent. From this you might reasonably conclude that the 347 should make about 13.4 percent more torque, less any that was lost to the extra piston friction from the longer stroke. An inspection of the

With the big-port Dart Pro 1 heads, the difference between 306 inches and 347 was far beyond what might normally be expected. A 13.4 percent increase in displacement delivered an average of 19 percent increase in output.

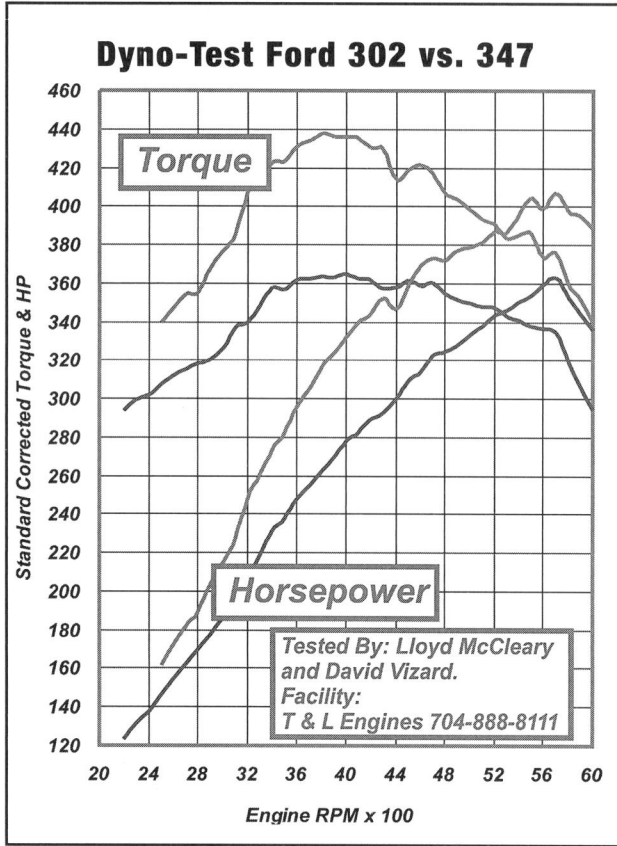

Dyno-Test Ford 302 vs. 347

Tested By: Lloyd McCleary and David Vizard. Facility: T & L Engines 704-888-8111

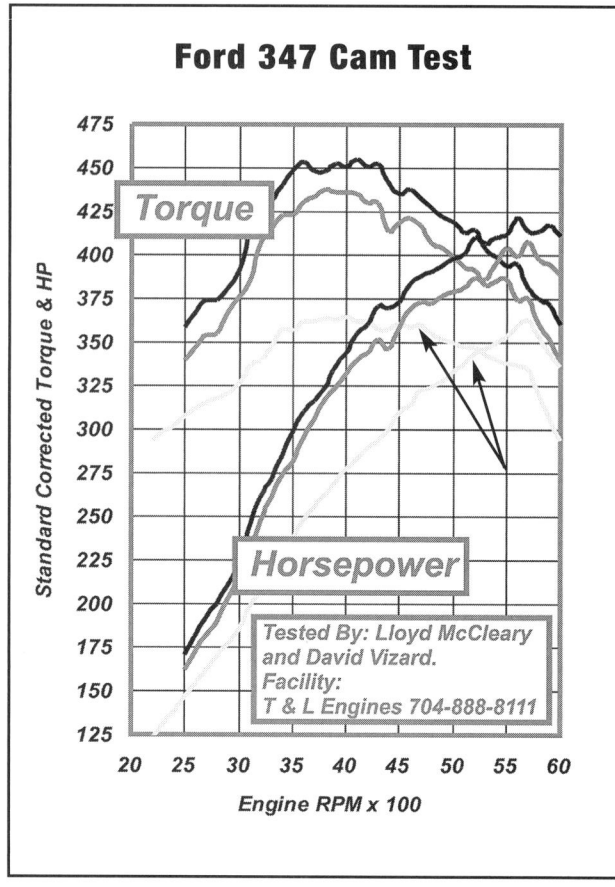

Ford 347 Cam Test

Tested By: Lloyd McCleary and David Vizard. Facility: T & L Engines 704-888-8111

numbers reveals that over the range tested, our Dart Pro 1–headed 347 stroker delivered an average 19 percent increase.

Not surprisingly, most peoples' reaction to that is, "How can that be?" Much as this may sound like a physical impossibility, it's not. In fact, it's just the start of a demonstration of how at the end of the day the right combination of parts—not the use of one or two apparently trick parts—really returns the dividends.

Remember that the original 195cc port Dart Pro 1 heads test premise was whether or not they could be successfully used on an existing 302-inch engine while the cash for a stroker kit was being rounded up? The 195cc port runner Pro 1s, though significantly more suited to a bigger-inch short-block, did deliver a sizable increase in output (52 hp) on a stock stroke, 0.030-over (for 306 inches) mule motor when they replaced the stock heads. But, as McCleary pointed out, a set of 170cc heads would have been far more suited to the smaller displacement, with the 195s being near ideal for hot 331 and 347 street motors.

Port velocity is much more important than the average hot rodder imagines. Having good flow along with a higher port velocity can considerably help cylinder filling after the piston has passed the BDC point of the intake cycle. Because the piston

Replacing a cam that was optimally spec'd for a 302 with one that was spec'd for a 347 produced an even stronger power curve from our 347. The cam, available only from T&L, was worth an average of 19 lb-ft more than the one it replaced, even though it used identical lobes. From this graph, it can be seen just how much the stroker kit with optimally spec'd parts is worth. The light-gray curves (arrows) are those of the original-baseline 302.

is now on its way up the bore, the only way to continue the filling process until the valve closes (in our case) some 74 degrees after BDC, is to convert the kinetic energy of the incoming air into pressure energy so it goes on pushing air into the cylinder. This function is dependent not just on the velocity, but also the square of the velocity.

Putting that into numbers, if the port velocity is increased by 10 percent, the ramming pressure goes up by 21 percent. This factor alone makes engines far more sensitive than you might think to port velocities. By utilizing such for extra cylinder filling, the torque output goes up throughout the rpm range. This means the engine reacts as if it were even bigger yet. As indicated by our tests, this in turn delivers more horsepower without an attendant increase in rpm.

So far, our stroker crank has paid off handsomely by delivering a hefty increase in torque everywhere in the rpm range, and that has translated into an equally hefty chunk of extra power. At the end of the day, the best power increase to have is by virtue of a torque increase everywhere in the rpm range, as it means that extra output is instantly on tap.

Though the results are good, up 'til now we should ask if there is more to this than just getting the port velocities in the right ballpark. The answer to that, as we shall see, is yes, as a big displacement increase has a measurable effect on the optimum valve events needed for the now-bigger engine.

Both our baseline 306 and the 347, as tested so far, had one of Comp Cams Xtreme Energy grinds (PN 35-421-8). McCleary's take on this cam was that within the confines of a cam with 276 degrees of off-the-seat duration and 500-odd-thousandths lift, you are unlikely to find any worthwhile increase no matter how long you dyno test—as long as it is a 302 with good heads you are testing it in.

Up the cubes to 347 and the picture changes. The valve event timing that was near optimal in a 302 now no longer applies. Since McCleary has invested a lot of expensive dyno time into establishing more optimal valve events for a 347, he was in no hurry to let the world at large know what the specs are for the cam he has developed.

OK, our 347 with its 13.4 percent increase in capacity is already up a near-amazing 19 percent in output. How much more can a cam deliver, especially one with the identical intake and exhaust lobes but different event timing? Well, you might be in for the second surprise here. Just check out the graph, page 110. The new cam was appreciably better everywhere in the rpm range. The cam itself delivered an average increase of 19 lb-ft increase over the original 302 Comp cam. That works out to be an average of 7.7 percent just for knowing what the combination needed. If you want one of these custom cams, they do cost a little more and can only be had from T&L.

Overall, our stroker combination was now up from the baseline output by an impressive 24.5 percent. What we have learned is that a stroker kit can produce excellent performance return per dollar spent, especially when it is used with the appropriate parts. If you want to make sure your stroker kit has those appropriate parts, then be aware that T&L can not only sell you a Scat/Ross stroker kit at a competitive price, but also the Dart heads along with an appropriate intake and carb plus the right cam for the job. If you don't want to build it, the company can do that for you, and, as an added bonus can dyno test it before shipping.

SOURCES

Advanced Engine Developments (AED)
2530 Willis Rd.
Richmond, VA 23237
804/271-9107
www.aedperformance.com

Barry Grant
Carburetors
1450 McDonald Rd.
Dahlonenga, GA. 30533
706/864-8544
www.barrygrant.com

Dart Machinery
353 Oliver St.
Troy, MI 48084
248/362-1188
www.dartheads.com

Edelbrock Corp.
2700 California St.
Torrance, CA 90503-3907
310/781-2222
www.edelbrock.com

KT Engine Developments
384 Industrial Ct.
Concord, NC 28025
704/784-2610
www.ktengines.com

Pertronix
440 E. Arrow Hwy.
San Dimas, CA 91773
909/599-5955
www.pertronix.com

Professional Products
12705 Van Ness Ave.
Hawthorne, CA 90250
323/779-2020
www.professional-products.com

Ross Racing Pistons
625 S. Douglas St.
El Segundo, CA 90245
800/392-7677
www.rosspistons.com

SCAT Enterprises Inc.
1400 Kingsdale Ave.
Redondo Beach CA 90278
310/370-5501
www.scatcrankshafts.com

T&L Engine Development
12303-A Renee Ford Rd.
Stanfield, NC 28163
704/888-8111
www.tandlengines.com

Chapter 15
HOOLIGAN HOT ROD 5.0

Going For the 9s With a 5.0

Text and Photos by the *MM&FF* Staff

The Hooligan Hot Rod lives. It's been a fun 2004 for mid 10s and all, but for now we're upping the ante and gunning for 9s.

There comes a time when Mustang ownership becomes a really, really bad obsession. Modification ideas and intentions start innocently in one direction and more often than not lead to an entirely different agenda until you just can't stop. One minute you're bolting on underdrive pulleys and the next you're at the chassis shop adding halo bars for 25.1C certification.

We say this because we purchased and built our '91 Mustang GT to be a homebuilt hot rod that would go 10s with a realistic budget, and that was it. Nothing more. Nada. We didn't expect the stock long-block 302/Lentech AOD combo to go as quickly as it did (10.60s at 128 mph). Not only did the car exceed our expectations, but it was quite a shock to everyone else involved. With only a turbo kit from HP, some great tuning from Mustang Magic, and a little home-brewed ingenuity, we walked away with super-quick times and no loss of driveability. With all the stock body panels still in place, the ultimate sleeper was born.

STAGE I: BOTTOM END

Now we've decided to go to the next level with our Wild Strawberry hatchback—to get it into the nines as a street car. We're not trying to prove any points here, since any Mustang techie worth their weight in PCV valves knows that a properly tuned turbocharged V-8 Mustang will go fast (read: very fast), and do it all day. Heck, most of you can probably build one that goes much faster than ours, but we're just ordinary guys, not experts. More importantly, we have

nothing to prove, only to write about.

In this stage we're going to upgrade the weak link in our nine-second chain, which is the stock mill. We'll put in a streetable engine combo that will make stupendous horsepower when the boost gauge is pinned, and we'll address the safety concerns with the car, since we're still running around with no rollcage and—even worse—stock

Want to build a stroker engine yourself, but not sure how to do it? Allow us to assist as we demonstrate the installation of Summit's low-buck 347 stroker kit.

We had a stock E7 block sent out to a local machine shop for some grindwork. The bores were opened up to 4.030 inches and the deck surfaces were nicely milled. Brass core plugs were then installed, and the entire casting was thoroughly cleaned. Here, we placed the block onto the engine stand after we washed it another time.

First off, flip the block upside down and insert the main bearing shells into the main saddles. Apply a light coat of oil onto both sides of the bearing. This ensures maximum heat transfer from the bearing surface to the block. The crankshaft is then carefully lowered into place, and some lube is placed on the crank's main journals. We use 80-weight gear oil as an alternative to break-in lube and get great results.

To get a good indication of how much bearing clearance there is, we used good ol' Plastigauge, which is included in the Summit stroker kit. Place a small piece of Plastigauge on each main journal of the crankshaft, install all five main caps with the bearings, and torque down the main bolts. Then, remove each cap and measure how much the small strip of plastic has spread. The wider it is, the more it was crushed and the less clearance is present. Recommended clearance is 0.002 inch with a variance of 0.0005 inch. We were spot-on with 0.002 inch for strong oil pressure and low rotating resistance.

The bottom of each bore requires a notch for the opposing cylinder's connecting rod bolt. With the careful use of a die grinder and a carbide cutting bit, you can get excellent results. The grinder setup is similar to the ones that head porters use to open up cylinder heads. The best way to properly make the notch is to install the crankshaft with the connecting rods in place to accurately remove the least amount of iron. This will keep as much material as possible on the bore for strength.

brakes in our rather portly (3,450-pound, with driver) GT. Even though the *MM&FF* staff could probably benefit from one less wacky editor, we think it would be nice to stick around for many more passes down the dragstrip.

Big-Armed Plan

The new engine will be a 347 stroker. Why a stroker, you ask? Simple. The most effective and reliable way to make horsepower in any application is to increase cubic-inch displacement. That's a no-brainer. But let's not forget the additional benefits from increasing the engine size. Stroking an engine allows you to keep the revs low to move air and build power, which means you won't need any high-tech (big-dollar) valvetrain equipment. This will also allow you to use factory-style hydraulic roller cams with stock lifters, and if lift is under 0.550 inch, conventional high-performance springs for 6,500-rpm applications can be used. This, in theory, will produce a torque-rich powerplant that is not high-strung, making it streetable, EEC-IV-friendly, and relatively bulletproof.

For the engine, we relied on Summit's affordable ($1,299) 347 stroker kit (PN CSUM347KIT), which comes with a Scat nodular iron crank, a set of bushed 4340 H-beam rods in 5.400-inch length, standard-sized Clevite77 rod and main bearings, a high-volume Melling oil pump, and a complete Fel-Pro gasket set. Normally, this kit comes with Keith Black hypereutectic pistons in stock 10:1 compression. Although the KB units work well in most performance applications, we did not think they would be ideal for a turbo-charged powerplant because of the compression ratio. Instead, we opted for Trick Flow Specialties' forged pistons, (also available from Summit), which are available in 8.5:1 compression and 4.030 bore size as PN 51404110.

The hardware was top-notch, and everything arrived within a few days. The only thing you'll need to remember is that the stroker kit is not balanced with either the KB pistons or the TFS slugs, so count on having the entire rotating

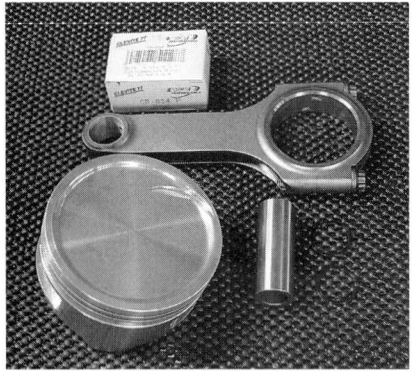

The Summit 347 stroker kit normally comes with Keith Black hypereutectic pistons that have 10.0:1 compression. Since that's way too much squeeze for our turbo'd combination, we upgraded to Trick Flow's dished turbo pistons for 347s. The forged slugs come with stock-sized 0.927-inch wristpins and spiral locking clips. The H-beam rods feature floating small ends with bronze bushings. It requires a performance rod bearing, which is Clevite 77 PN CB634P (included).

First, phase the rod properly. We recommend doing one bank of pistons at a time, so that the rods on cylinders 1 through 4 have the chamfered ends facing the front of the engine, and that cylinders 5 through 8 have the chamfered ends facing the rear of the engine. With light oil on the wristpin, gently insert it into the piston and bushed end of the connecting rod.

With a little oil and a flathead screwdriver, carefully work your way around until the clip is fully seated. Repeat for the other side. It's a pain, but be patient so that you don't hurt yourself. This is a good time to use eye protection.

On the big end of any connecting rod in a V-type engine, you will find one side with a large chamfer on it. This is because when two rods are next to each other, oil seeps away from the middle and to the outside of the rod journal. This is known as rod phase, and it is important to keep the chamfered end facing outward on each crank throw.

Next up are the pesky spiral clips. These are usually made of high-strength, high-carbon spring steel, so they can be a bear to work with. We recommend first pulling and bending them apart (as shown) to open up the clip. This will make it a lot easier to insert the clips into the receiver grooves of the piston.

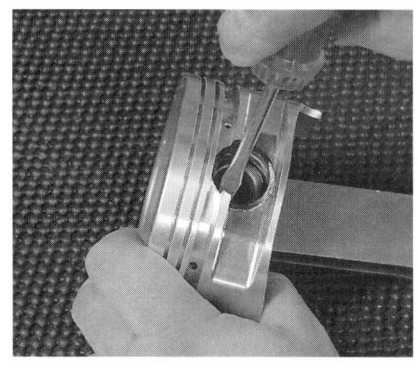

With all the rods and pistons assembled, now is the time to prepare the piston rings. With the top and second rings squared into each bore (you can use one of the pistons and rods for this), you'll want about 0.020–0.022 inch of end gap. Many engine builders have their preferences for end-gap clearances, but we've had excellent results on street cars with power adders at 0.020 inch. To get the desired gap, we place the ring into a vise with soft jaws and use a fine, flat file with even strokes. Each pass of the file removes approximately 0.001 inch of material.

assembly gone over by the local machine shop. Frankly, we'd do it anyway for our own benefit, even if it was advertised as a balanced assembly.

When using a stroker kit, keep in mind that most kits are not based on the flyweight 50-ounce imbalance crankshaft. This means your factory harmonic damper and flywheel/flexplate will not work with any 347 stroker kit. You will need new ones that are for a 28.2-ounce imbalance. They can vary in price, but quality increases or decreases accordingly. Not wanting any compromises, we elected to use TCI's Rattler damper (PN 870007) for the utmost in harmonic absorption and its SFI-approved flexplate (PN 529628). With the rotating assembly gathered, we brought everything—including the stock block—to a local machine shop for $750 worth of cutting and pasting. In the end, we walked away with a completely balanced rotating assembly and a block that was notched, decked, bored, and honed 0.030 inch over.

And yes, we did say "stock block." We know the limits of the E7 casting are rather mediocre, at best, but thanks to the aftermarket, we can continue using the stock casting for power levels that are more than twice what the factory had intended. As a matter of fact, we've witnessed several 600-rwhp combinations with the factory lump. The key ingredient to making it a reality is a quality main girdle. For this, we turned to Trick Flow once again for its billet-steel piece (PN 51500700) that works

Because the oil ring intersects the wristpin bore, a hardened steel spacer is required to retain the bottom of the oil control ring. It simply goes into the groove and sits down low. Keep the open end of this ring away from the pin ends.

With the oil ring pack in place, move onto the lower compression ring. Note that the small dot on the ring indicates this is the top side. With a ring compressor, expand the ring and place it into the middle groove. The top ring goes on next, so keep in mind which side goes upward.

Ring phase is another thing to look for. We like to have the top ring face the front of the engine and the second ring face the rear. This makes the open ends 180 degrees away from each other, minimizing the chances of cylinder pressure to have a straight shot into the crankcase.

The rod bearing halves each have a tab that locks into receiving grooves in the connecting rod. With light oil (10W-40 works fine) lube the backside of the shell, and slide it into place so that it looks like this. Remember when we were talking about rod phase? In this picture, you can see that the chamfered side also makes the bearing sit offset toward the adjoining connecting rod.

Although we've used other types of ring compressors in the past, experience has proven that the tapered designs are the easiest and safest to use. Just make sure you get one for your exact bore size. With the pistons and rods all ready for action, we slid them into their new homes.

for stock stroke engines as well as those with 3.400-stroke cranks like our 347.

It also comes with ARP main cap bolts, so it saves you additional parts hunting. Thankfully, the 1/2-inch-thick main girdle does not require machining of the main caps for installation, which saves time and money (both of which we often have very little of).

The camshaft we elected to use was Trick Flow's smallest grind, 51402000. With 220/225 degrees at 0.050 inch lift and 0.499/.510 total lift at the valve, the specs are healthy but far from over the top. Knowing that a turbo car doesn't need as much cam as a radical, naturally aspirated engine would, we looked to keep it on the mild side since torque production down low was one of the key traits we wanted to keep. In maintaining a strong torque curve, the spool up of the turbo is enhanced and driveability is still kept. We could have gone with a more radical bumpstick, but unless you're building an all-out race engine, too much duration and less overlap literally just blows boost out of the tailpipes. Turbocharged street cars are where you don't want to get too wild with the cam selection.

Click-type torque wrenches are more convenient than the traditional beam-types, but some argue the old-school designs are more accurate. Whichever wrench you use, tighten the rod bolts to the recommended 63 ft-lb.

The Trick Flow main girdle was designed to not require any machining of the main caps and to accommodate 347-inch stroker engines. After removing the stock main cap bolts, we torqued down the supplied ARP main bolts to 80 ft-lb. The factory specifications call for 70 ft-lb, but with high-quality fasteners that are more resistant to stretching (higher tensile strength) you can up the torque value for increased clamping capacity.

The last thing to check is crankshaft endplay. In order to do this, you need to loosen the center cap (third journal) and tap it around until you get the desired endplay of 0.004–0.008 inch. We kept it in the tight side at 0.004 and torqued the main cap bolts back to spec.

Even with new oil pumps, we always like to remove the cover and check for anything fishy inside. Always remove the pump gears and check for burrs or debris. This high-volume Melling pump was perfect inside and out.

The oil-pump driveshaft included is a step up from the spindly stock unit. However, with the increased oiling needs of a high-performance engine—especially one with an external power adder that relies on engine oil for lubrication such as a turbo or supercharge—we opted for a super-strong Ford Racing unit. PN M-6605-B302 works with all fuel-injected '86–'95 302s.

The oil pickup tube's mounting tab needs to be modified to work with the main girdle. You'll have to cut it shorter by about 0.75 inch and drill a new hole to line up with the girdle that is tapped in this location for the included 1/4-inch bolt.

Up next is the careful installation of the cam. On turbocharged street cars, you don't want to go too crazy because the cam's profile has a strong correlation to how quickly a turbo spools up. To keep this engine torquey, we went with a mild TFS grind with 220/225 degrees of duration at 0.050 lift and 0.499/0.510 lift on the intake and exhaust, respectively. To install a hydraulic roller camshaft, simply slather it with motor oil and slide it in.

Place the camshaft retaining plate as shown and lift it slightly so that it is centered on the cam. Tighten the two bolts to 10 ft-lb and use thread locking compound.

Note the use of threaded oil galley plugs rather than simple press-in plugs. This ensures a positive seal that will not work loose due to vibration. We followed up with the installation of the Summit double-row timing chain set at the straight up position (no advance or retard).

With a light dose of motor oil, insert each hydraulic roller lifter with the orifice facing upward. This will make any air bubbles inside of the lifter rise and bleed out. This will prevent lifters from losing pressure overnight, which causes the dreaded lifter tick whenever you start the engine the next day.

The lifter guide plates (dog-bones) and retainer were then installed. With the rotating assembly finished, we took a gander at our new mill and felt almost sad that we had to cover it up with more engine parts.

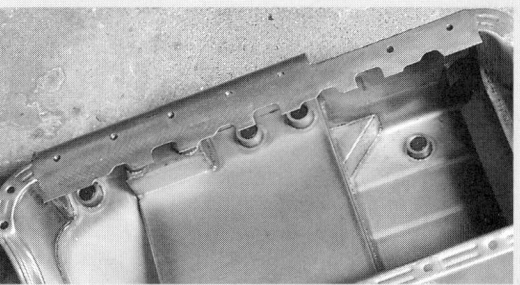

We ordered a deep-sump Canton oil pan from After Hours Racing, since the company had one in stock and was nearby in New Jersey. The pan has an integral crank scraper that helps control oil as it flies off the crankshaft. It works for stock-stroke engines out of the box, but it needed clearancing for our 347.

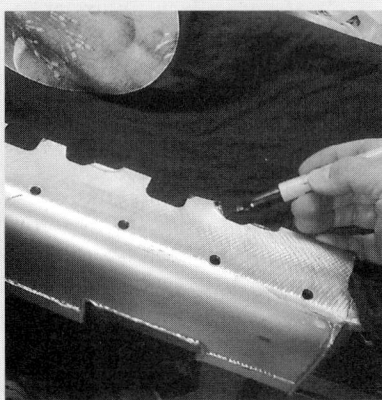

We placed the pan into position and marked off the points of contact with the rotating assembly, which was pretty much everywhere.

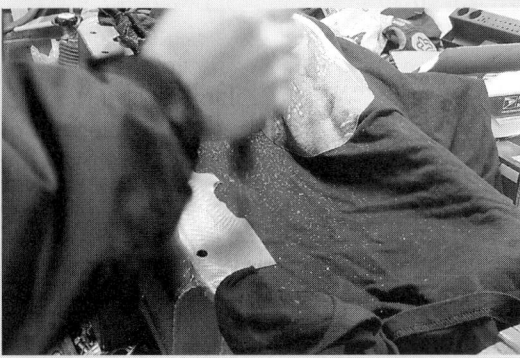

With the help of a die grinder and a carbide cutting tip, we made short work of the pan. Once done, we cleaned it out thoroughly with brake-cleaner spray.

We trial fit the pan one more time to make sure the oil pump pickup tube had sufficient clearance with the bottom of the pan. Ideally, you'll want between 7/16- and 1/2-inch of an air gap. You can use modeling clay or even tape a bolt to it to see how much space there is. This will ensure adequate oil flow. With the pan taken care of, we installed the front timing cover and bolted the pan into its final place. Next, we'll pay attention to the topside of things as we turn this short-block into a long-block and install it into the Hooligan Hot Rod.

To keep our cylinder pressure contained, we relied on a set of Cometic laminated-steel head gaskets that we sourced from Bennett Racing. The advantages are superior cylinder sealing without the added expense of special machine work.

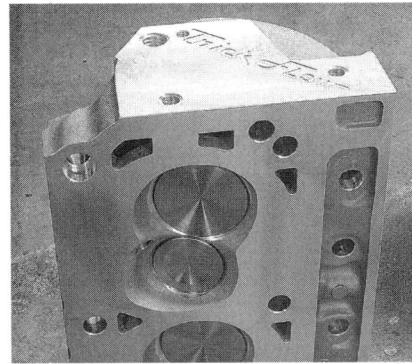

As part of the Trick Flow top-half engine package for the fuel-injected 302, the aluminum Twisted Wedge cylinder head is a street-worthy piece that does not sacrifice usable low-end torque for high-end performance. With our turbo application, we found its 170cc intake runner and 61cc chamber to be ideal for our moderate-boost (15-psi) engine. Right out of the box, the assemblies come with 2.02/1.60 valves and single valvesprings designed for hydraulic roller cams. All we had to do was wipe them clean and bolt them onto the block.

STAGE II: TOP END

It's time to pay some attention to the meat and potatoes of horsepower production, which is the top end of our engine. Rather than go with a gaggle of race-only engine components, we elected to use a more driveable and user-friendly package: Trick Flow's top-half kit (PN K514-350-370), which includes its Twisted Wedge aluminum cylinder heads, Street Heat EFI long-runner intake manifold, mild hydraulic roller cam, cast-aluminum valve covers in silver finish, and 1.6 roller rockers.

Miscellaneous parts to complete the install are included and consist of ARP head bolts, a Fel-Pro gasket set, hardened chrome-moly pushrods in 6.700-inch length, and a double roller timing chain set. The all-inclusive kit allows you to build an engine entirely out of a box, using only your old intake manifold bolts.

Although we could have traveled down a more radical route, we elected to build an engine based entirely on street-type parts that you might buy if you walked into this hobby for the first time. Aside from retaining the car's streetability, we wanted to keep build costs down. Since the complete TFS top-half kit retails for $2,075, it doesn't break the bank, yet still produces gobs of usable torque and horsepower for the street. In addition, it avoids many of those late-night runs to the local speed shop for parts that would be needed had we put this combo together in a piecemeal fashion.

One thing that most long-time Mustang enthusiasts forget is how great "street-type" parts are. The "bigger is better" ideology (or dare we say, "idiotology") is a dangerous one—a credit card in one hand and a phone in the other can cause you to fall into the pit of overkill. This not only drains your wallet of hard-earned cash, but ofttimes creates a situation where parts are not matched for their respective performance levels. It

To prep the deck surface, use a strong solvent and a lint-free cloth rag to remove any oil. We used brake cleaner on the rag rather than spraying the surface and then wiping it down. This prevents accidentally spraying the cylinder walls with the solvent, which would remove the oil from the bores that lubricate the pistons upon initial startup.

With the head dowel pins in place, carefully lay the gasket into position. You don't want to scratch any of the special coating off the gaskets.

often results in a final product that is actually worse than if they simply had ordered a more conservative list of components.

Thankfully, the Trick Flow top-half package comes as a matched kit, so the guesswork is eliminated. Although you won't build a 600hp naturally aspirated engine with these components, you will build one that is fully reliable and tractable enough for day-to-day driving. Best of all, this keeps the fun quotient high and makes your car a joy (rather than an irritable nuisance) to own.

And, no, don't think we are salaried by Trick Flow for telling you this. Our advice holds true for any other manufacturer of hard-core performance components such as Edelbrock, Brodix, or World Products. They're the ones spending hundreds of hours on the dyno with specialized engineers so that all you have to do is make a phone call to enjoy

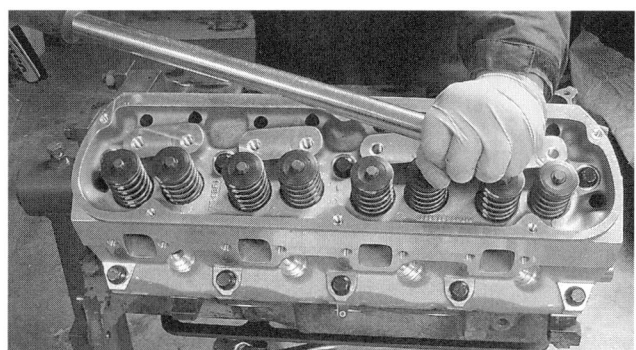

Lower the head onto the block, making sure the dowel pins are fully seated into the head. With the head bolts lubricated, we torqued all the bolts down to 90 lb-ft of torque.

We installed the supplied rocker arm studs and guideplates, and slid the new Trick Flow pushrods into their final locations.

The Trick Flow roller rockers are also part of the Trick Flow top-end kit, and are adjusted by doing the following: After making sure the lifter is completely down, tighten the poly-lock by hand until pushrod contact is made. From here, use a 5/8-inch wrench and tighten the poly-lock one full turn. Then, bottom out the small hex grub screw and tighten the poly lock about 1/8 of a turn. Do the exhaust valve when the intake is about to close, and set the intake valve when the exhaust is about to open.

Trick Flow supplies these neat cast-aluminum valve covers that provide clearance for the roller rockers underneath. Although not included, we opted for OEM Ford valve cover gaskets that came on '91-and-up Mustangs. These are stamped steel with a rubber perimeter seal and require no sealant, working well over and over again.

the fruits of their labor. Besides, there's nothing nicer than seeing a brown truck come down the street with all your new goodies in it. It's like Christmas any time of the year.

KEEPING THE PRESSURE WITHIN

Although there have been many leaps in cylinder-sealing technology—including O-ringed decks, O-ringed heads, receiver groove head gaskets, and numerous composite and steel combinations for improved seal—we've only recently stumbled across possibly the ultimate solution: laminated-steel head gaskets. For decades, high-performance engine builders and even some OEMs have used laminated steel gaskets in their engines for the utmost in durability on high-horsepower applications.

The gaskets compress evenly when mated to freshly machined surfaces, promoting good deck seal for the cylinder pressures and the cooling transfer passages. They go on dry without any special sealants and require no machining whatsoever. The only difficulty is, getting them can

The Fel-Pro intake manifold gasket comes with a tab that locks into a conventional head gasket to keep it in place while you are installing the intake manifold. However, the Cometic head gaskets do not have a tang that traditional composite gaskets do, so you'll have to use silicone sealant to keep the gasket in place and maneuver it into final position while installing the intake. Also note that a small part of the gasket needs to be cut from the coolant transfer ports to prevent a restriction.

be a pain at any local parts depot. Fortunately for us, Jon Bennett had a pair of Cometic-branded gaskets in stock. Because this recent hot commodity has been a top seller for him and his company, Bennett Racing, he has plenty on his shelves. We got our parts from Mr. B., and within a few days we were ready to start building some serious power.

ASSEMBLY LUBE REQUIRED

As we mentioned, the hard part is now behind us as we assembled the bottom end last month to the exacting standards of a coal shoveler. All our clearances were well within range, and the engine appears to turn freely. Once we buttoned up everything, we put the fresh 347 into our beautifully unbeautiful '91 GT and drove the car about 100 miles before we changed the oil. We then loaded up the trailer, took our beloved Hooligan Hot Rod to Old Bridge Township Raceway Park in New Jersey, and were rewarded with slightly better times. Keep in mind that our previous best with the stock 302 was a 10.690 at 127.47 mph on a cool, crisp autumn afternoon. The best pass with our new engine was 10.31 at 133.2 mph. We gained well over three tenths but picked up over 5 mph. It may not seem all that impressive, but consider this—it was done without any dyno tuning and with only 120 miles on the engine.

Another factor was the 94-degree day. Had we been fortunate enough to have had a 68-degree day like we did last year, we feel this would have sent this car flying into the 10.0 range. Now, we'll head over to the dyno and see how much more we can extract. With the weather cooling off, we're aiming for nines and think we have everything in order with our hardware. We just need to get off our lazy butts and make it happen.

SOURCES

After Hours Racing
16 S. Michigan Ave.
Kenilworth, NJ 07033
908/241-4773
www.afterhoursracing.com

Bennett Racing
1640 11th Ave.
Haleyville, AL 35565
205/486-5520
www.bennettracing.com

Edelbrock
2700 California St.
Torrance, CA 90503
800/416-8628
www.edelbrock.com

Summit Racing Equipment
P.O. Box 909
Akron, OH 44309
800/230-3030
www.summitracing.com

TCI Automotive
151 Industrial Dr.
Ashland, MS 38603
662/224-8972
www.tciauto.com

Trick Flow Specialties
1248 Southeast Ave.
Tallmadge, OH 44278
330/630-1555
www.trickflow.com

A nice, thick bead of silicone RTV sealant works wonders to seal the ends of the intake. Lay down a smooth bead like this, and cap off the engine.

With a quick layer of paint applied, we removed the engine from the stand and installed the rear oil galley plugs that were previously inaccessible. They are 1/4-inch NPT threads. This is also the time to install the rear main seal and to make sure the engine-to-transmission dowel pins are secure and in place.

With the factory block-plate in place (always reinstall this piece because it spaces the starter, torque converter, and crankshaft properly), we then bolted up TCI's SFI-approved flexplate for our application. When using an AOD behind a 347, you will need a 164-tooth ring gear with a 28-ounce imbalance. Late-model 302s require a 50-ounce imbalance.

With the engine neatly nestled into place, we installed an Edelbrock Victor Series water pump and began hooking up our front-drive accessories.

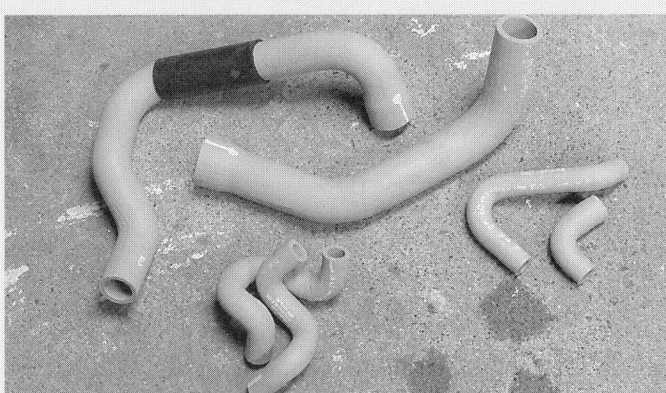

With the new engine comes more power, but more power usually means more heat. To combat any potential cooling problems, we installed a Ford Racing silicone hose kit from After Hours Racing.

To keep the temperatures constant, we installed a Robertshaw–balanced 180-degree performance thermostat into the factory water neck.

On the compressor side, the HP Performance 66mm turbo (left) is only slightly larger than the 60mm unit we removed (right), according to the naked eye. But a closer look reveals a compressor wheel of an entirely different shape and design to offer increased airflow and better response.

The 60mm turbo (left) that we had on the 302 before was considered too small for the new engine by the folks at HP Performance. To better match the characteristics of the new 5.6-liter engine, we opted for a lightly larger 66mm unit (right). The turbine housing was also upped from a tight 0.81 A/R to a larger 0.96 A/R to maintain quick spoolup and improved exhaust flow.

Turbocharger turbine housings play a vital role in the turbo's response and overall power output. This was the old 0.81 housing that we removed. Looking outside, we could not tell any difference. But closer inspection of the larger 0.96 housing (not shown) revealed a wider exhaust nozzle that reduces exhaust backpressure at the expense of exhaust gas velocity that would hit the turbine wheel.

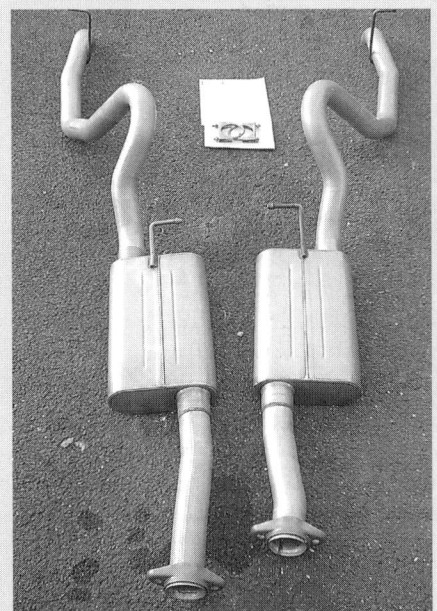

To complement the new engine, we finally ditched the old exhaust, which was a bad combination of crush-bent, small-diameter pipes and no-name mufflers. We opted for Edelbrock's RPM Series after-cat exhaust kit that offers full 2.5-inch mandrel-bent piping and chambered mufflers that offer a deep tone without excessive noise. We like things quiet on this car.

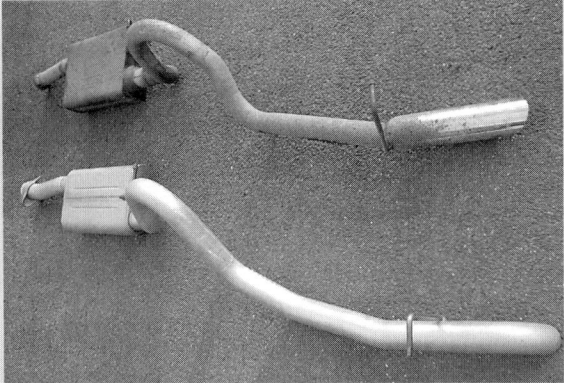

Compared to our stock tailpipes and cheesy exhaust tips, the Edelbrock kit looks like a million bucks. From the looks of things, it outflows the old exhaust considerably. It's hard to believe we went 10.60s with the old tubing.

To make the Edelbrock exhaust mate to our HP Performance turbo kit's mid-pipe, we cut a few inches off the flowtubes and welded the pipes straight on. An electric reciprocating saw makes short work of this otherwise elbow-busting job. We used a 115-volt at-home Mig welder and achieved excellent results in our installation. With the car now complete, we will dyno tune our new engine combo and see if we have enough power to go nines.

Chapter 16

TWIN SHERMANATORS

A 530hp Pump Gas Stroker and a 475hp All-Out Racer

Text and Photos by Luke Magnus

SHERMANATOR I: THE STREET MOTOR

Joe Sherman Racing in Santa Ana, California, is one of those "boutique" engine shops that cater to local racers and street guys who have a real need for speed. It doesn't matter if you bring him a Chevrolet, AMC, Oldsmobile or Ford, Sherman relies on his trusty flowbench, dyno and 40-plus years of experience to put together a winning combination.

Unlike many engine builders, Sherman has a thorough understanding of the complete racecar package. He and his sons have raced incredibly quick bracket cars for years. Sherman knows from first-hand experience how much power you need, and where in the powerband you need it. Wringing every last ounce of useable power out of a given combination is his specialty; in fact, he won the first inaugural Popular Hot Rodding Engine Masters Challenge—besting pro engine builders from across the country. So when the "Engine Master" turned his attention to a pair of 347ci stroker Fords, *MM&FF* wanted to be on hand to record the experience.

The plan was to build two 347s. Shermanator I was designed to run on pump gas, make 500 hp and not break the bank. Shermanator II will be a flat-out, single four-barrel race setup. To keep the confusion to a minimum, this section will cover the buildup and test of the pump gas street/bracket motor. In the next section we will show how Sherman made 750-plus hp and some very impressive torque numbers from his 347 race mill.

Initially the pump gas 347 was to be built around a budget rotating assembly. Joe ordered the parts and when the pistons arrived with several different styles of valve notches, Joe

Joe selected the lightest and strongest piston he could find. This JE flat top forging weighs just 414 grams. The Chevy-style pin adds another 105 grams. Rods are Scat 4340 I-beam-style that are 5.40 long.

decided to revert to Plan B, using engine components he knew and trusted. It would cost a little more, but knowing the engine would stand up to several seasons of 7,000 rpm abuse would more than make up for the difference in price. Joe is definitely on the frugal side, but he spends money where it counts—on anything that enhances reliability and power output.

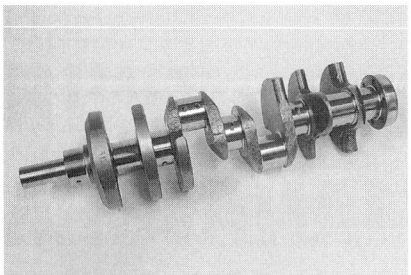

Sherman swears by Scat's crankshafts. This Scat 9000 cast crank features radiused counterweights and chamfered oil holes and comes ready for assembly.

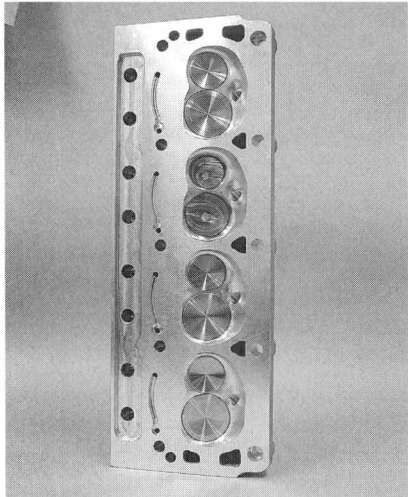

Airflow Research 185 aluminum cylinder heads were selected for this combination. They feature 58cc combustion chambers that set final compression at a pump gas–friendly 10.6:1.

Swirl-polished, stainless steel 2.02 and 1.6 valves have undercut stems and are part of the AFR head package.

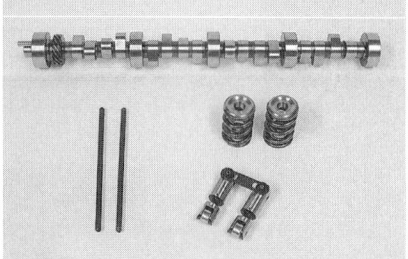

Roller cam and lifters are from Comp Cams. Joe selected one of their latest designs with 248/252 degrees duration at .050 and .662- and .648-inches of lift with Jesel 1.6 shaft rockers. It's ground on a 108-degree centerline. Valvesprings are part of the AFR 8000 head package.

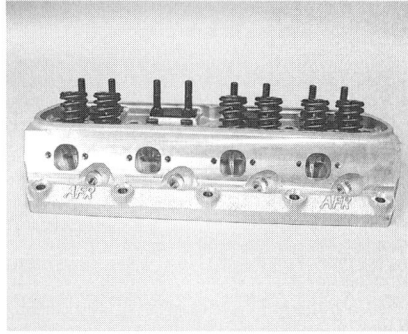

The exhaust ports look a little puny, but the flowbench and dyno say they're just right. Joe claims that he is able to improve flow slightly with a little grinding.

The block is a late-model production 302. The reason we mention this is that later blocks have slightly different deck heights than early blocks. Joe's block ended up with an 8.18-inch deck height after final machining. After inspecting the O.E. main bearing caps he made the observation that they looked a little "delicate," so he installed a CAT main bearing girdle that requires no machining. It helps retain the Scat 9000 cast crank, which features radiused counterweights and chamfered oil holes. Joe has used a number of Scat cranks and swears by their quality, durability and price. The crank rotates in Federal Mogul Competition Series bearings with clearances kept on the tight side of recommended tolerance.

Scat got the call for the

connecting rods as well. Joe selected a set of 4340 I-Beam-style rods that measure 5.40 inches center-to-center. The longer rods (stock measure 5.09 inches) allow the pistons to clear the stroker crankshaft counterweights and provide a more favorable rod ratio. These rods come with ARP bolts and are bushed and drilled on the pin end for oiling.

One of Joe's "speed secrets" is that he always uses the lightest and strongest piston available, because it takes power to stop and accelerate a piston twice per crankshaft revolution. The JE forging (PN 188703) fills Joe's requirements perfectly. They feature a flat-top design and weigh a mere 414 grams, plus another 105 grams for the Chevy-style premium piston pin. A set of 1/16-, 2/16- and 3/16-inch file-fit Speed Pro rings provide the ring seal and glide-on cylinder walls that are finished with an 820-grit stone. The oil ring is a low-tension design, but has enough oil control for a street engine without the assist of a vacuum pump. The remainder of the bottom end consists of a stock pressure and volume Mellings M-68 oil pump and a Stef's welded aluminum, high-capacity oil pan.

Joe is constantly doing R&D on camshaft profiles to find more horsepower, however he's really pleased with Comp Cams' latest series of solid roller lifter profiles. The cam he selected for this pump gas 347 measures 248 and 252 degrees duration at .050-inches of lift, and .662- and .648-inches of lift at the valve with 1.6-ratio Jesel shaft rockers. It's ground on a 108-degree centerline and installed in the motor straight up.

Due to the aggressive profile on this cam, Joe feels that the effective duration is more than the specs would indicate. In fact, the dyno tests revealed that this engine combination could use a little less cam timing, because the actual compression is only 10.5:1. When the valve lash was increased (decreasing duration), it picked up the power.

Jesel Sportsman Series shaft rockers are an affordable alternative to studs, roller rockers, guideplates and girdles. They reduce friction at the valve tip, stabilize the valvetrain and maintain lash longer.

Professional Products offers a budget series of race manifolds, in satin and polished finishes. Joe selected the single-plane Hurricane model that is rated for 3,500–8,000 rpm.

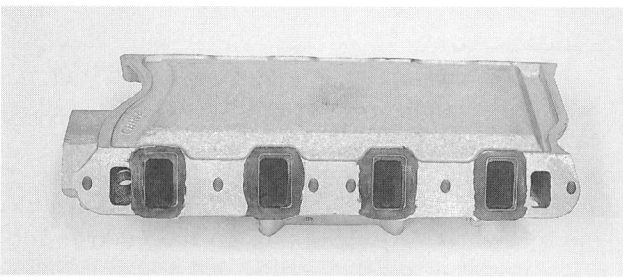

Joe port-matched the intake to the heads. The Professional Products matched up with the AFR 185s pretty well, but won't work on the larger AFR heads.

Jesel shaft rockers may sound extravagant for a pump-gas street engine, but not when you take a closer look. Jesel's Sportsman Series (SS) shaft rockers retail for just $695, which is no more than you would pay for a set of good aftermarket studs, roller rockers, guideplates and stud girdle. Plus you get all of the advantages of shaft rockers, such as better geometry from the rocker's longer pivot length, the stability of the shaft that ties it all together and high-strength adjusters that maintain valve lash longer.

The real key to making 530 hp from 347 ci on pump gas is the Airflow Research (AFR) aluminum cylinder heads. AFR is selling these heads like hotcakes. They come in a few different sizes and options. Joe selected AFR 185s for this engine. They feature 2.02/1.60-inch stainless steel valves and 58cc combustion chambers that set the compression ratio at 10.6:1 with the .053-inch Cometic head gasket Joe is running (more on that later).

Joe ordered the optional AFR 8000 roller cam spring package that includes springs set at 210 pounds on the seat and 550 pounds open. They will accommodate valve lifts to .670 inches and come with titanium valvespring retainers. Joe slapped them on the flowbench and performed some minor porting. He's amazed at how good these heads are out of the box. He said that you can make slight improvements in flow, maybe 5–10 cfm, but the average guy will most likely hurt the flow numbers, so leave them alone. The heads are held in place by ARP hardware.

Things got interesting when Joe got to the 347's intake system. He heard that there were some excellent components on the market that would make the power numbers he wanted and save a few bucks in the process. He selected a Professional Products Hurricane single-plane manifold that is available in a satin or polished finish. It has a range of 3,500–8,000 rpm and has some neat features such as integral bosses that can be drilled and tapped for nitrous nozzles. Joe went in and smoothed the port dividers in the plenum area. He said that he does this regardless of whose manifold he uses; the Hurricane just needed a little more grinding than usual to meet his approval.

The carburetor choice was a big surprise. Joe contacted Quick Fuel Technology (QFT) in Bowling Green, Kentucky. He gave them his engine specs and application and let them design a carburetor to match the application. No wimpy 750 here! QFT provided one of its Q-Series 950-cfm 4150s. They start with a ProForm carb body, bore the venturis out to 1.450 inches, install double-step downleg boosters and a billet baseplate with 1 3/4-inch throttle blades. The throttle shafts are "slabbed" and use button-head screws to improve flow. On the metering side, QFT two-circuit billet metering blocks are installed with four-stage emulsion. Other features include notched floats, jet extensions and

Regardless of what manifold he uses, Joe reworks the plenum and runner divider area for more flow. He claimed the Professional Products intake required more grinding than most, but the results were excellent.

Quick Fuel Technology (QFT) supplied a Q-Series 4150-style carb rated at a true 950 cfm. It features an enlarged ProForm carb body, billet metering blocks, jet extensions, notched floats and double-step downleg boosters.

The cool thing about QFT's Q-Series is that they are a true race carb with vacuum ports that allow them to be run on the street. Check out the billet baseplate, 1 3/4-inch throttle blades and "slabbed" throttle shafts.

Other details include a Professional Products SFI-approved, harmonic balancer and a Stef's aluminum oil pan. Distributor is a Ford housing with an Ignitioneering mag pickup assembly.

Shermanator II is just your typical, everyday 751hp, 357 small-block buildup.

high-flow squirter screws. QFT's Q-Series carbs are streetable with timed vacuum ports, PCV ports and full vacuum ports. However, they don't have a choke. With a $639 retail price the Q-Series is approximately $300 less than a competitive custom race carb.

Joe assembled the 347 with Fel Pro gaskets and ARP fasteners. Everything went together as planned, except the pistons ended up .010-inch above the deck. Joe hadn't realized that he was using an early block with the shorter deck height. No problem! His ace-in-the-hole was the Cometic Multi Layer Steel (MLS) gasket that is available in thicknesses for .027 to .120 inches. The .054-inch-thick gasket was just the ticket to establishing the correct clearances. Other details include a Ford distributor setup with an Ignitioneering mag pickup trigger mechanism, an MSD ignition box and SVO valve covers adding some style.

Joe hooked the 347 to his trusty Superflow pump, installed 1 3/4-inch dyno headers and 3-inch Magnaflow mufflers. He threw in a set of ACCEL 784 spark plugs gapped at .045 inches and set the timing for 38-degrees total. A set of #78 primary and #86 secondary jets were selected (after a few warm-up pulls) to meter the 91-octane Unocal gas. Joe made more than 14 dyno pulls in all, optimizing ignition timing, valve lash and jetting. The final result was 527.0 hp at 7,000 rpm, and 443 lb-ft of torque at 5,600 rpm. Torque never dropped below 400 lb-ft from 4,800–6,800 rpm. (Joe experimented with a larger 1050-cfm QFT carb and larger cam and produced 530.2 hp at 7,000 rpm, but slightly less torque.)

What Joe Sherman has put together here is a pump gas small-block that has an ideal power curve on a budget that most can relate to.

SHERMANATOR II: THE RACE MOTOR

When Joe first proposed building two 347 stroker engines for *MM&FF*, he knew the race motor would actually be a 3.5-inch stroke 357, but he neglected to tell us—at least until his SuperFlow dyno rocked to the tune of 751 hp at 8,400 rpm.

"The power was still climbing at 8,400, but because my dyno console sits right beside the engine, I didn't have the nerve to rev it any higher," Sherman says. "Heck, with a sheetmetal tunnel ram, this thing might make 800 hp." Concerning the deception about the increased stroke, the cagey Sherman casually explained it this way. "If it made typical 347 power, I wasn't going to mention the longer stroke."

Once a street racer, always a street racer.

What's remarkable about this engine combination is its simplicity. No power adders, just a single four-barrel, mildly ported cylinder heads, plenty of off-

The forged 4340 Scat 3.5-inch stroke crank weighs just 43 pounds and comes with windage-reducing aerodynamic counterweights and polished journals with chamfered oil holes. Sherman just cleans it and puts it in the block.

CP 4.030 bore two-ring pistons weigh just 400 grams. The Manley 2 1/2-inch-long, 0.927-inch-diameter pins add 78 grams. Manley 5.40-inch-long aluminum rods add another 420 grams. Calico applied antifriction coating to the piston skirts and the Federal-Mogul bearings, along with an oil-shedding coating to the rods.

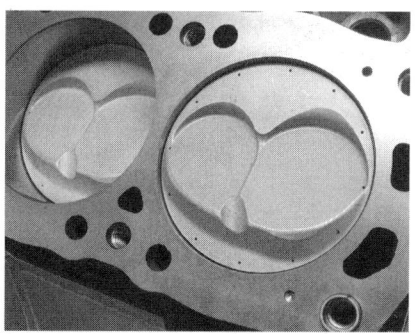

Piston domes were cut with huge valve reliefs and drilled for 0.040-inch gas ports. Domes were coated with a Calico thermal barrier. The block is a Ford Racing 8.200-inch deck version that can be bored to 4.125. It comes with billet four-bolt main caps and a wall-thickness sonic-inspection sheet.

Here you can see the beautiful billet four-bolt main caps and the extra meat in the skirt area of the Ford Racing block. It weighs almost 50 pounds more than a production passenger block.

the-shelf components, and a lot of Sherman's savvy. The real genius was putting together the combination. Joe attributes much of the engine's power output to two things—reduction of drag and a low reciprocating mass. From the outset, the engine was designed to use a two-ring piston with low tension rings that require a vacuum pump to provide oil control and improve ring seal.

Additionally, anything that could be coated to reduce friction was. Extremely lightweight pistons, pins, and rods were used to improve power. It takes power to stop and accelerate the piston twice for every revolution of the crank. With a low piston/rod mass, the counterweights on the crankshaft can be lighter, allowing the engine to accelerate faster. Many people overlook these simple physics of building a performance engine, but Sherman knows that you take every horsepower you can get, wherever you can get it. So let's see what it takes to build a 2.1hp/inch, single four-barrel, small-block Ford.

INSIDE THE BELLY OF THE BEAST

Sherman based this 357-cid beast on a Ford Racing 8.200-inch-deck block. This is as close to a custom aftermarket race block as any O.E. will ever build. It comes with a bore size slightly under 4.00 inches, so it can be finished to 4.00 inches or overbored a whopping 0.125 inch. The block comes complete with a signed and dated sonic inspection report. The cylinder walls in this particular block were 0.350 inch thick with virtually no core shift. The three center four-bolt main caps are machined out of billet, and the main webs and skirt area around the bores are much thicker. In fact, the block weighs almost 50 pounds more than a standard 302 block. If you are looking to run a longer stroke or a

piston with more compression height, Ford Racing offers a similar block with a taller 8.700 deck.

To arrive at the 357 mark with the 3.5-inch stroke crank, Sherman had the block bored and finished to 4.030. CP made the custom gas-ported, two-ring pistons. Sherman has CP make all his custom pistons because of the company's five-axis CNC mills that can lighten pistons in ways most manufacturers cannot. The domes were cut for huge valve reliefs and drilled with 0.040-inch gas ports. They were also machined for 0.043-inch compression and 3mm oil rings. They weigh just 400 grams and are fitted with Manley 78-gram tool steel pins that are 0.927 inch in diameter, 2 1/2 inches long, and have a 0.090-inch wall thickness. Sherman sent the pistons to Calico Coatings for an added antifriction coating to the skirts and a thermal barrier to the tops.

Manley supplied the 5.40-inch-long aluminum connecting rods that weigh just 420 grams. They were coated with a Calico coating designed to shed oil. We asked Sherman about the useable lifespan of aluminum rods and if they could be run successfully in a street engine. He seemed to think that 450 runs in an automatic drag car was reasonable, and that they could be used in a dual purpose street/strip car with no problem. He recalled one aluminum rod maker installing a set in the shop truck and logging 100,000-plus miles.

A 3.5-inch-stroke Scat lightweight (43 pounds), forged 4340 crankshaft rounds out the rotating assembly. It has aero-style "boat tail" counterweights that reduce windage at high rpm. These cranks are so finely finished (including chamfered oil holes and polished journals) that Sherman simply gives them a

Comp Cams supplied the huge solid roller lifter cam and lifters. It features 278 and 285 degrees of duration at 0.050 with 0.789 inch of lift. It is ground on a 112-degree centerline that produced a broad, flat torque curve from 5,000 to 8,400 rpm.

The front of the motor is all business with a Jesel beltdrive that provides precise cam timing that is easily adjustable ± 10 degrees in just minutes. The beltdrive also allows quick cam changes through the front cover. The ATI Super Damper is covered up with an MSD crank trigger ring.

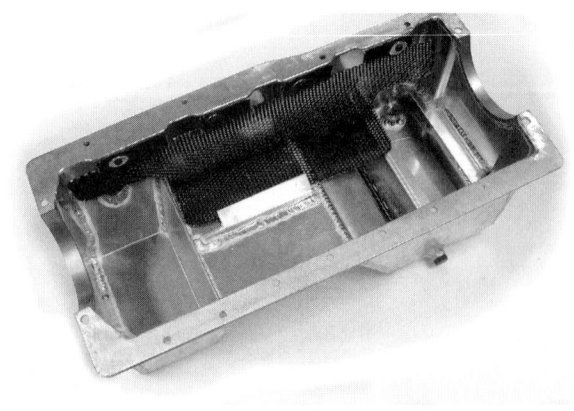

The Stef's billet pan is a piece of work. It is engineered to strip oil off the rotating assembly and return it to the pan's sump without splash. It gets filled with 10W30 Pennzoil.

final wash in the solvent tank before dropping them into the Federal-Mogul bearings. An ATI harmonic damper was used to ensure longevity and safety at extreme rpm. The bottom end was wrapped up with a Stef's fabricated aluminum oil pan and pickup and a Mellings oil pump set at 58–60 psi.

Like the previous street 347, Sherman selected a set of aluminum Air Flow Research heads. He used AFR 185s on the street engine, and AFR 205s for this motor. They come with 58cc combustion chambers. He used a 0.040-inch Cometic MLS head gasket that set the final compression ratio at 13.25. AFR makes a larger 225 version of this head, but Sherman feels they are more at home on larger-displacement or supercharged engines. He spent hours on the flow bench and with his grinder to pick up about 10 cfm per port, but cautions that it is easier to lose flow if you don't know what you're doing. "These heads will make great power right out of the box, so leave them alone," he says. He also installed Ferrea 2.08- and 1.60 stainless steel valves after having both intake and exhaust ports coated with a thermal barrier coating. The valvesprings are Manley's Nextec race springs that are unique in that they only have three-and-a-half to four coils, which eliminates the possibility of coil bind in all but the largest cams. They were set up with 275 pounds of seat pressure and 750 pounds at 0.800 inch of valve opening.

The remainder of the valvetrain consists of a huge Comp Cams solid roller cam that specs out at 278 and 285 degrees of duration at 0.050 and carries 0.789 inch of lift.

It is ground on a 112-degree centerline. Remarkably, this cam produced a relatively broad torque curve with 452 lb-ft at 5,000 rpm, peak torque of 502 at 7,200 rpm, and barely trailing off to 470 at 8,400 rpm. Comp Cams roller lifters team up with Jesel SS 1.6-ratio shaft rockers to produce a valvetrain capable of 9,000 rpm. A Jesel beltdrive is used for quick cam timing changes. It also dampens out crank-to-cam harmonics and aids in Sherman's campaign to reduce friction wherever possible. The beltdrive features a billet-aluminum front cover that was designed to accept electric water pumps. In this case, a Meziere electric pump is used.

GOING TOPSIDE

Moving to the induction system, Sherman selected an Edelbrock Super Victor intake manifold. Its ports closely match the AFR heads, and in past tests Sherman has been pleased with the power output. He did the basic port matching, then went into the plenum area and rounded and smoothed the edges of the port dividers. He stresses that the plenum work is time well spent. For some reason, the air/fuel mixture does not like sharp-edged port dividers. The manifold is topped with a 2-inch Super Sucker combo spacer that starts out as a four-hole spacer under the carb, then transitions into an open spacer within 1 inch.

Speaking of carbs, a Quick Fuel Technology 4150-style carburetor rated at 1,050 cfm was selected to handle the fuel. It features a Proform main body with bores enlarged to 1.590 inches. The billet-aluminum baseplate holds 1 3/4-inch throttle blades. It features a tuneable three-circuit metering system with screw-in bleeds for the idle, intermediate, and high-speed circuits. The reverse-flow intermediate circuit provides instant throttle response in applications like this that have a weak booster signal.

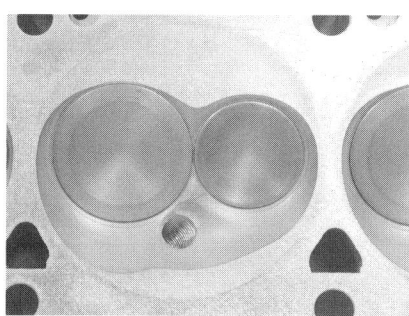

Air Flow Research 205 heads proved to be the perfect match for Sherman's 357 engine combination. Here you can see the 2.08 and 1.60 Ferrea stainless steel valves, and tiny 58cc combustion chambers coated with a Calico heat barrier coating.

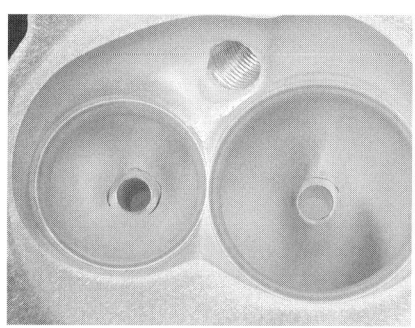

Calico also coated both the intake and the exhaust ports to isolate heat from the intake charge and to retain heat and velocity on the exhaust side.

The assembled Air Flow 205 head shows Manley's Nextec Race valvesprings that have only three-and-a-half to four coils, which eliminates the possibility of coil bind even with the Comp 0.800-inch-lift cam. Fel-Pro gaskets provide the intake seal.

The Super Victor intake was port-matched to the cylinder heads, and Sherman did some minor cleanup work in the plenum area. It made impressive power over a wide rpm band.

A Quick Fuel Technology 1,050-cfm, 4150-style carb was used with a 2-inch-thick Super Sucker combo spacer. It features a three-circuit metering system with a reverse-flow intermediate circuit that provides excellent throttle response for applications with low booster signals.

Hidden under the MSD crank trigger ignition setup is the ATI Rattler 2000 damper. The crank trigger connects to an MSD Digital 7 box. A Meziere electric water pump bolts directly to the Jesel beltdrive, and the GZ Motorsports vacuum pump is driven from an underdrive crank pulley. The engine was built around the GZ vacuum pump that added close to 50 hp.

In spite of all of this tuning capability, Sherman found the carb to be perfectly calibrated right out of the box.

Distributor chores were left to a stock Ford housing with an Ignitioneering Magnetic Trigger conversion hooked to an MSD Digital 7 box. An MSD crank trigger setup is used for precise cylinder firing. Total advance was set at 35 degrees for the VP 115 race gas. NGK No. 9 plugs were used for this dyno test.

One of the final components bolted on the engine was one of the most important—the GZ Motorsports vacuum pump. It is beltdriven off the crankshaft and pulls vacuum from the valve covers. Without it, the low/no-tension piston ring setup wouldn't be feasible. In fact, Sherman dynoed the engine with and without the pump, and the difference was 50 hp. But before you bolt one on your engine with a traditional ring setup expecting that amount of gain, you should know that Sherman has tested vacuum pumps on several con-ventional engines, with results ranging from 0 to 15 hp. Also, be aware that the vacuum pump requires a catch can, and, depending upon your air/oil separator, it can pull quite a bit of oil out of the engine.

Some of the final details that went into the buildup included ARP fasteners, Fel-Pro gaskets, a few quarts of 10W30 Pennzoil, and a K&N Gold oil filter.

Finally, it was reckoning time with the SuperFlow dyno. Sherman's usual ritual after engine break-in is to make a few part-throttle partial load pulls to check the fuel and spark calibration. Those numbers were in the ballpark, so he set the dyno to sample at 600 rpm/sec and began the dyno pull. The power numbers were incredible. It started with 431 hp at 5,000 rpm, reached 504 at 5,600, 608 at 6,500, climbed to 706 at 7400, and marched right on up to 751 at 8,400. This is one mean, single four-barrel small-block!

There may have been a few horsepower left on the table above 8,400 rpm, but like we said earlier, Sherman thought 750 hp was enough, though he would like to try for 800 hp with a sheetmetal intake and a pair of carbs.

SOURCES

Air Flow Research
10490 Ilex Ave.
Pacoima, CA 91331
818/890-0616

ARP
1863 Eastman Ave.
Ventura, CA 93003
800/826-3045
www.arp-bolts.com

ATI Performance Products
6747 Whitestone Rd.
Baltimore, MD 21207
800/284-3433
www.atiperformanceproducts.com

Calico Coatings
6400 Denver Industrial Park Rd.
Denver, NC 28037
www.calicocoatings.com

Cometic Gasket
8090 Auburn Rd.
Concord, OH 44077
440/354-0777
www.cometic.com

Comp Cams
3406 Democrat Rd.
Memphis, TN 38118
901/795-2400
www.compcams.com

CP Pistons
1902 McGaw Ave.
Irvine, CA 92614
949/567-9000
www.cppistons.com

Edelbrock
2700 California St.
Torrance, CA 90503
310/781-2222
Tech Only: 800/416-8628
www.edelbrock.com

Federal-Mogul Performance
26555 Northwestern Hwy.
Southfield, MI 48034
248/354-2700
www.federal-mogul.com

Ferrea Racing Components
2600 NW 55th Ct. #238
Fort Lauderdale, FL 33309
888/733-2505
www.ferrea.com

Ford Racing Performance Parts
15021 S. Commerce Dr., Ste. 200
Dearborn, MI 48120
800/367-3788
www.fordracingparts.com

GZ Motorsports
22338 Shake Ridge Rd.
Volcano, CA 95689
209/296-3793
www.gzmotorsports.com

JE Pistons
15312 Commercial Ln.
Huntington Beach, CA 92649
714/898-9763
www.jepistons.com

Jesel
1985 Cedarbridge Ave.
Lakewood, NJ 08701
732/901-1800
www.jesel.com

Joe Sherman Racing
2302 W. Second St.
Santa Ana, CA 92703
714/542-0515
www.joeshermanracing.com

Manley Performance Products Inc.
1960 Swathmore Ave.
Lakewood, NJ 08701
732/905-3366
www.manleyperformance.com

Meziere Enterprises
220 S. Hale Ave.
Escondido, CA. 92029
800/208-1755
www.meziere.com

MSD Ignition
1490 Henry Brennan Dr.
El Paso, TX 79936
915/855-7123
www.msdignition.com

Professional Products
12705 S. Van Ness Ave.
Hawthorne, CA 90250
323/754-1287
www.professional-products.com

Quick Fuel Technology
2352 Russellville Rd.
Bowling Green, KY 42101
270/793-0900
www.quickfueltechnology.com

SCAT Enterprises Inc.
1400 Kingsdale Ave.
Redondo Beach, CA 90278
310/370-5501
www.scatcrankshafts.com

Stef's Fabrication Specialties
693 Cross St.
Lakewood, NJ 08701
732/367-8700
www.stefs.com

When it comes to Ford modular engines, is bigger really better?

Chapter 17

STOKED ON STROKE

Studying the Effect of Displacement on the Dyno

Text and Photos by Richard Holdener

When looking for ways to improve the power output of your mod motor, one of the avenues open to you is increased displacement. This should be especially important to those of you considering replacing that tired stock engine you've managed to somehow keep alive despite the abuse you've subjected it to over these long years.

If you think your stock 4.6 is in need of an overhaul or you're looking to step up to a more stout combination to better withstand the rigors of boost, you'd do well to consider a stroker combination. We stepped up to a stroker (similar to the one tested here) for Project Ice Box and were rewarded with impressive power gains on the supercharged GT.

The benefit of a stroker is that instead of building a short-block that will withstand the rigors of forced induction, nitrous, or just the daily grind, the short-block will actually be responsible for improving the power output. Just like the ported heads, cams, and even the blower you installed, the bottom end will actually increase power production not just at the peak but throughout the rev range. Such is the benefit of a stroker short-block.

To illustrate the effect that an increase in displacement has on the power curve, we had to build not one but two different mod motor combinations. Number one was a traditional Two-Valve 4.6 consisting of a stock crank combined with aftermarket forged connecting rods and forged aluminum pistons. The pistons were chosen to produce a static compression ratio of 10.8:1 with our CNC-ported PI heads. We chose the elevated compression ratio to further enhance power production. The 4.6 short-block came from Coast High Performance.

The 4.6 short-block consisted of a steel Cobra crank equipped with forged rods and pistons to produce a static compression ratio of 10.8:1.

The forged short-block was topped off with a set of CNC-ported heads from Total Engine airflow. The CNC porting dramatically enhanced the flow rate of the PI heads, allowing us to take maximum advantage of the large XE278 AH Xtreme Energy cams from the Comp Cams catalog. The XE278 AH cams offered 0.550 lift and a 242/246 duration split. The cams were installed with new lifters (lash adjusters), while the valvespring package supplied by Total Engine Airflow for its CNC-ported heads allowed us to rev right past 6,000 rpm without fear of valve float. This 4.6 combo was

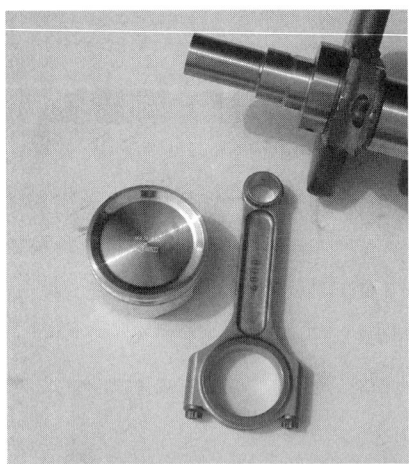

Coast High Performance supplied the 5.0 modular stroker kit. Similar to the one used on Project Ice Box, the increase in displacement came from a slight overbore (0.020) and a healthy increase in stroke length (from 3.54 inches to 3.75 inches).

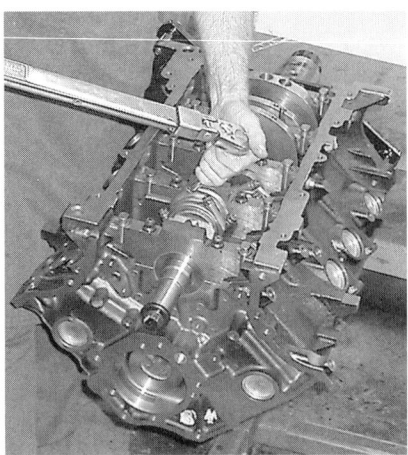

CHP even went to the trouble of assembling the 5.0 stroker assembly, to which we had to add the TEA heads, cams, and PI intake.

The 4.6 and 5.0 were treated to a set of Stage 3 CNC-ported heads from Total Engine Airflow. The TEA heads offered extensive porting, oversized Manley valves, and even combustion chamber polishing.

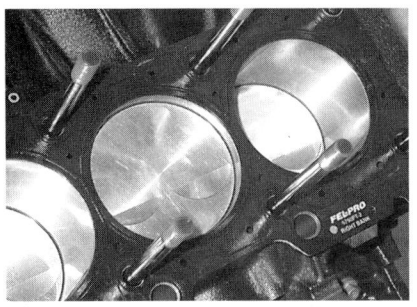

Both test mules were assembled using ARP head studs and Fel-Pro MLS gaskets

also fitted with a stock PI intake, though we did augment the performance with an Accufab 75mm throttle body and inlet elbow. Finishing off the 4.6 was a set of Kooks 1 5/8-inch long-tube headers, a Meziere electric water pump, and the FAST management system.

The 4.6 was warmed up, broken in (for 25 minutes), and finally run in anger on the engine dyno. It produced impressive power, especially when considering it displaced only 281 ci. How many old-school, street-oriented 289 small-blocks do you know that thumped out nearly 400 lb-ft of torque? Tuned to produce an air/fuel ratio of 13.0:1 with 28 degrees of total timing, the 4.6 produced 405 hp and 394 lb-ft of torque.

The 4.6 produced peak power at 6,000 rpm, while the peak torque occurred at 4,800 rpm. Torque production exceeded 375 lb-ft from 4,200 rpm to 5,500 rpm, and exceeded 350 lb-ft from 3,750 rpm to 6,100 rpm. The 4.6 exceeded 350 hp from 4,700 rpm all the way to 6,200 rpm, where maximum acceleration will be taking place. We've run a number of Four-Valve 4.6s that did not produce this much power, so it's nice to know that despite the lack of a couple valves per cylinder, the Two-Valves motors can be made to produce some pretty impressive power.

Next up was the 5.0 version. The heart of the 5.0 stroker from Coast High Performance was the

stroker crank. Where the stroke on the stock 4.6 crank measured 3.543 inches, the CHP 5.0 stroker kit upped this measurement to 3.75 inches. In addition to the increase in stroke, the CHP 5.0 also featured a minor increase in bore size of 0.020 over (from 3.552 to 3.572). It is possible to go as large as 0.070 over to achieve 5.1 liters, but the 5.0 designation holds a special place in our hearts. The 5.0 kit featured a new 3.75-inch crank, a set of forged connecting rods, and forged aluminum pistons. The kit also came with the necessary piston rings and bearings, though we elected to test our combination in assembled short-block form. Assembled by CHP, the 5.0 was topped by the same set of TEA CNC-ported cylinder heads, PI intake, and Comp XE278AH cams. We also employed the same Kooks 1 5/8-inch stainless steel headers, Meziere electric water pump, and FAST management system controlling the 36-pound injectors. In short, the 5.0 was equipped with the same components used on the smaller 4.6, offering only an increase in displacement.

What we expected and received from the increase in displacement was an increase in power. The CHP stroker produced 434 hp at a slightly lower 5,800 rpm and 430 lb-ft of torque at the same 4,800 rpm. Torque production on the larger stroker motor exceeded 400 lb-ft from 3,750 rpm to 5,150 rpm. Even at 3,500 rpm, the stroker thumped out 375 lb-ft of torque, a number carried all the way to 6,000 rpm.

While the stroker easily outperformed the smaller 4.6, it is interesting to note that the 4.6 actually bettered the larger stroker in terms of specific output (defined as power output per displacement). The 4.6

Comp supplied the necessary Xtreme Energy cams for both motors. In this case, we selected a set of XE278AH grinds that offered 0.550 lift and a 242/246 duration split.

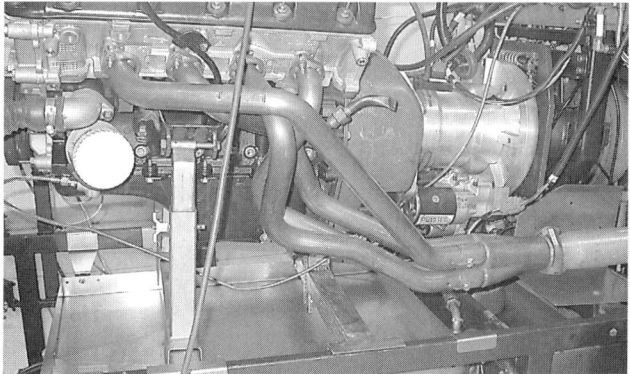

We relied on a set of 1 5/8-inch Kooks stainless steel headers for the dyno testing.

The stock composite PI intake was run on both engines. In retrospect, it probably hurt the stroker more than the 4.6 in terms of maximum power.

The PI intake was upgraded with an Accufab 75mm throttle body and matching inlet elbow. The combination was worth a good 17 hp over the stock components on the stroker.

Run with the FAST engine-management system, the modified 4.6 produced 405 hp at 6,100 rpm and 395 lb-ft of torque at 4,800 rpm.

Using the same components, the larger 5.0 stroker motor produced 434 hp at 5,800 rpm and 431 lb-ft at 4,800 rpm. Note that the peak power occurred 200 rpm lower on the stroker, yet the two produced peak torque at the same engine speed. This might be the sign of an intake restriction.

produced 405 hp, giving it a specific output of 88.04 hp per liter, while the more powerful 5.0 was down at 86.8 hp per liter (434 hp/5.0L).

An examination of the power curves indicated that the stock PI intake manifold may have been the restriction on the more powerful 5.0, as it was obviously never intended to supply the airflow needs of an extra 170 hp (260 hp in stock trim vs. 434 hp). Despite the need for a revised intake design, the stroker showed a consistent improvement over the 4.6 throughout the rev range. If you're thinking about stepping up to a new short-block, the extra cubes offered by the CHP stroker will definitely come in handy.

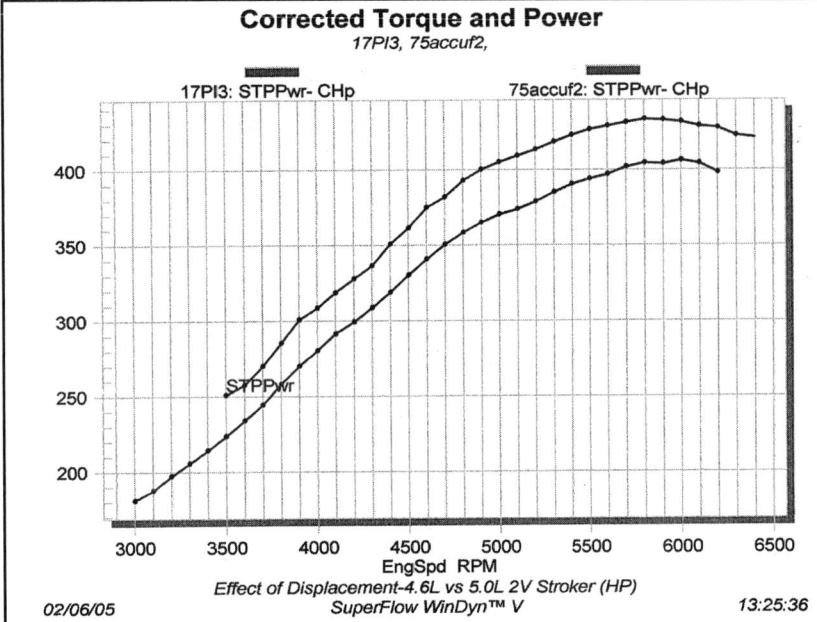

Corrected Torque and Power
17PI3, 75accuf2,

17PI3: STPPwr- CHp   75accuf2: STPPwr- CHp

Effect of Displacement-4.6L vs 5.0L 2V Stroker (HP)
SuperFlow WinDyn™ V

02/06/05   13:25:36

Effect of Displacement: 4.6 vs. 5.0 (Horsepower)
Stepping up from a 4.6 to a 5.0 had a positive effect on the power curve. Note the consistent power gain across the rev range. The interesting thing is that the displacement was the only thing to change—the heads, cams, intake, and headers were all kept constant, as was the static compression ratio. Oddly enough, the specific output calculated by the peak power numbers generated give the edge to the smaller 4.6, as the 405hp 4.6 (standard bore) produced 88.04 hp per liter, while the larger 434hp 5.0 checked in at just 86.8 hp per liter. It is possible the factory PI intake could have been restricting the power output of the larger stroker combination.

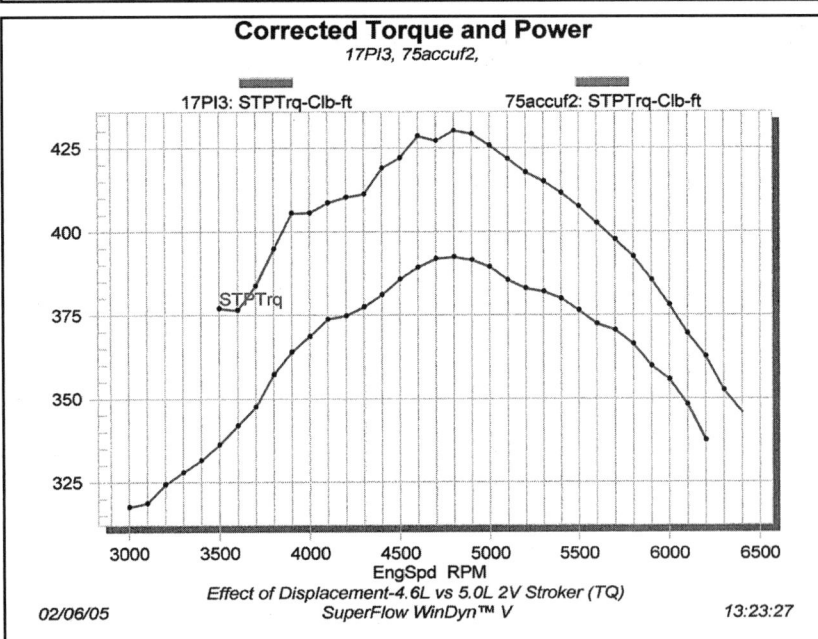

Corrected Torque and Power
17PI3, 75accuf2,

17PI3: STPTrq-Clb-ft   75accuf2: STPTrq-Clb-ft

Effect of Displacement-4.6L vs 5.0L 2V Stroker (TQ)
SuperFlow WinDyn™ V

02/06/05   13:23:27

Effect of Displacement: 4.6 vs. 5.0 (Torque)
The torque production offered by the larger stroker motor exceeded 425 lb-ft, while the smaller 4.6 struggled to reach 400 lb-ft. Note that the torque curves converged out near 6,000 rpm—a sure indication that the 5.0 was experiencing some type of restriction. Mathematically speaking, a consistent horsepower gain across the rev range will produce a diverging torque curve, but not to this extent. Note that the stroker did manage to exceed 400 lb-ft of torque from 3,850 rpm to 5,650 rpm.

SOURCES

Accufab
1516 E. Francis
Ontario, CA 91761
909/930-1751
www.accufabracing.com

Coast High Performance
2555 W. 237th St.
Torrance, CA 90505
310/784-1010
www.coasthigh.com

Comp Cams
3406 Democrat Rd.
Memphis, TN 38118
901/795-2400
www.compcams.com

Fuel Air Spark
Technology (FAST)
3406 Democrat Rd.
Memphis, TN 38118
901/260-FAST (3278)
www.fuelairspark.com

Kooks Custom Headers
59 Cleveland Ave.
N. Bayshore, NY 11706
866/586-5665
www.kookscustomheaders.com

Total Engine Airflow
525 Kennedy Rd.
Akron, OH 44305
330/784-5210
www.totalengineairflow.com

Chapter 18

INCHES & POUNDS

**Building a Turbo Motor
Capable of 200 MPH**

Text and Photos by Richard Holdener

The recipe for real 200-mph performance is pretty straightforward. Take one aerodynamic vehicle and add the necessary power to help overcome the aerodynamic resistance. Obviously, the greater the resistance, the greater the required power. Fox-chassis Mustangs have never been recognized for their slippery exterior, though there are worse designs on the road. In fact, the cars are probably most famous for their impressive power-to-weight ratio and the accompanying acceleration.

While the rest of the 5.0 world seems content to make short, quick passes down a narrow stretch of dragstrip, I have always danced to the beat of a different drummer. As far back as I can remember, the very first question I had relating to any automobile was always, how fast will it go? Whether the car in question was a hot exotic, a lowly airport rental or even a 20-year-old Ford Festiva, the question was always the same. While you might think the answer was self-explanatory, oftentimes additional discussion became necessary, especially to those raised on drag racing.

How fast will it go does not mean the trap speed measured at the end of the quarter-mile, but rather how fast will it go if you had sufficient room to leg the motor all the way up against an aerodynamic wall or the available gearing/engine speed? I can tell I've lost a great many when they get the look that questions why on earth you would want to keep going after the end of the quarter-mile? Others seem to grasp the idea (seemingly for the first time) and I can see the wheels beginning to turn as they mentally calculate the available power and gearing. It is these few disciples that eventually wind up running out in the Silver State Open Road Race. After sampling the intoxicating elixir known as top speed, most never look at drag racing the same way again.

Now that I've alienated the vast majority of the readership, I can tell you that drag racing and top speed (or open road

TOP—The D.S.S. 408-stroker short-block included forged dished pistons to keep the compression reasonable on our turbo motor. Note the ARP heads studs and Fel Pro 1011-2 head gasket. BOTTOM—In addition to the 4.0-inch steel stroker crank, the D.S.S. short-block featured 6.2-inch forged connecting rods and a Ford Racing Sportsman block.

D.S.S. also supplied a billet main stud girdle and windage tray. Also visible in this shot is the high-volume oil pump and Canton pick up.

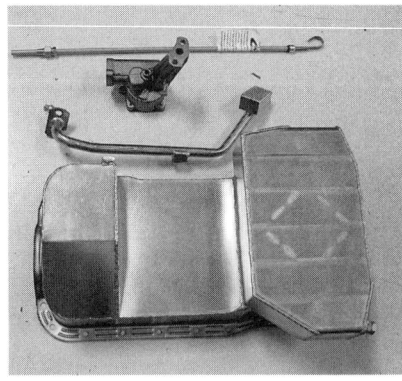

Canton supplied the road race oil pan along with the pick shown previously. Note the custom dipstick used with the Canton pan.

The Canton pan featured trap doors to keep oil in constant contact with the pick up.

Comp Cams supplied the XR286R Street Roller camshaft. The XR286R cam offered a .614/.621 (net) lift combo, a 248/254 duration combo all ground on a 110-degree lobe separation angle. The cam worked so well when tested on the 331-turbo motor, we decided to employ it on this larger 408.

Air Flow Research came through with our head gear, a set of AFR 205s.

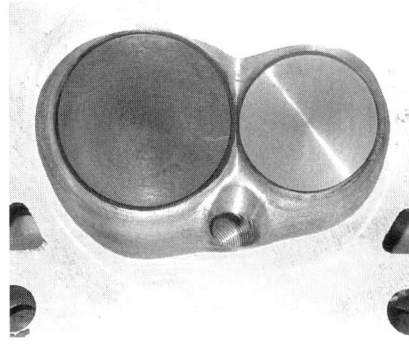

The AFR 205 featured a 2.05/1.60 valve combination along with (measured) 58cc combustion chambers.

racing) actually have more in common than I've let on. Naturally power production is at the forefront of both endeavors, but the similarities are actually more acute, in that the power production should actually be rpm (or application) specific for maximum performance. Given specific gear splits, final drive gearing and weight, it is most effective to produce maximum power in the range most used during the quarter-mile run. That is to say it is desirable to trade power production in an rpm range not used during the run to maximize power production in the usable rpm range.

If the gearing (splits and final drive) dictates that the engine be run from 4,500 to 6,500 rpm, then power production at 2,000 rpm is really irrelevant. The motor should be equipped with the appropriate cam, intake and cylinder heads to build the maximum amount of power in that range with less concern for power production elsewhere (somewhat less true with a daily driver).

Maximum top speed requires the very same concern for the operating range. This is especially true of the most critical shift (fourth to fifth or fifth to sixth) depending on transmission. The reason for the single-gear concern is in almost all applications, there is sufficient power to pull to redline in the lower gears and the power production only becomes critical when attempting to pull the final gear spread.

Let's suppose we have the same gear split as the drag race car and that shifting from fourth to fifth dropped the engine speed from 6,500 to 4,500 rpm. Suppose again our top-speed Mustang runs 155 mph at 6,500 rpm in fourth and now must not only be able to sustain 155 mph at the power produced at 4,500 rpm, but hopefully accelerate back up to the power peak which is probably much closer to 6,500 than 4,500 rpm. Thus the power production is critical from 4,500 to 6,500 rpm just like the Mustang running down the strip. The difference is the drag racer has geared his car to top out at 120–130 mph, while the top-speed Stang will be looking at a much greater aerodynamic load by passing to the high side of 155 mph.

It is with this in mind that I selected this particular motor combination for the 200-mph Mustang. Building power can be done in many ways. My '88 Mustang has enjoyed a dozen or so top speed runs out in Nevada running a mild 302 equipped with a Vortech supercharger. That combination produced top speeds in excess of 190 mph, but I never managed to actually see an honest (radar verified) 200 mph. I've seen nitrous used on top-speed vehicles as well, but nitrous should be used in small doses (actually time allotments) and I do not like to stop to have the horsepower bottle filled.

HP Performance supplied one of its turbo kits for use on our 408. Similar to the kit run previously on the smaller 302-based 331, the major difference was the length of the crossover tube.

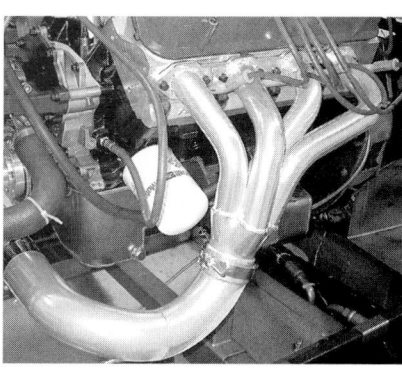

The entire turbo kit from HP Performance featured Jet-Hot coating to improve both the longevity and good looks. Check out the trick 4:1 header used to maximize exhaust energy to the turbo and the use of V-band clamps.

The AFR 205s featured an upgraded spring package and a set of Crane Gold 1.6-ratio roller rockers for use with our Street Roller cam.

In normally aspirated guise, the 408 was equipped with an Edelbrock Victor Jr. and a Holley 950 HP carb. Note also the MSD billet distributor, cap and wires.

Before adding the HP turbo kit, the 408 was run in normally aspirated form. The 408 produced 562 hp and 545 lb-ft of torque. Despite the single-plane intake and healthy roller cam, the 408 still managed to thump out 439 lb-ft of torque at 2,500 rpm.

Wanting something different, but equally effective, I chose to retain the use of forced induction, but decided to switch to turbocharging in place of my centrifugal supercharger. The largest benefit offered by the turbocharged motor was a broad power band. There is no denying a centrifugal supercharger has the potential to produce exceptional power, but so too does a turbo. Unlike the supercharger, the turbo produces a non-linear boost curve, meaning that maximum boost does not come near redline. The boost pressure curve is determined by a number of variables, but in most cases the boost is controlled by the wastegate once maximum boost is achieved. The benefit of the turbo is maximum boost can be reached long before the same boost level is achieved on a centrifugal supercharger.

Naturally having more boost earlier in the rev range has a positive effect on the shape of the torque curve. In direct comparisons (at a given boost level) a properly sized turbo may offer an extra 100 lb-ft of torque in the mid range compared to the same motor equipped with a centrifugal supercharger. It is this extra torque production that may pay huge dividends when it comes time to pull from 4,500 rpm in fifth gear against a 155-mph aerodynamic load.

If we assume we have two motors (one supercharged and one turbocharged) of equal peak power ratings (say 600 hp), we can surmise that both have the ability to produce the same top speed in the same vehicle (aerodynamics, gearing, etc.). Though both have the same peak power numbers, the vehicle equipped with the centrifugal supercharger may never get to reach the speed offered by the power peak, as the power present from 4,500 rpm may be insufficient to allow the car

The 408 was run sans air-to-air intercooler on the engine dyno due to space constraints. Since we were running just 7 psi of boost, the intercooler was not mandatory, but will be run in the car.

Without our fuel injection available, we had to run the turbo 408 with a carburetor. Once again we chose the Paxton carb bonnet to enclose the Mighty Demon carburetor.

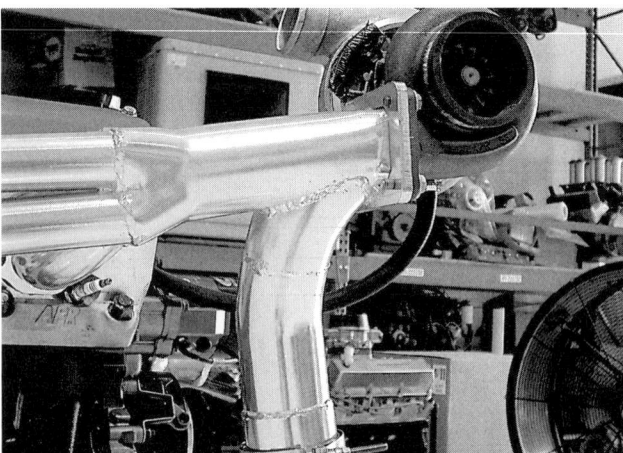

The turbo kit featured this free-flowing Y pipe to connect the passenger side exhaust manifold to the crossover pipe. It is attention to detail like this that makes the HP kit an exceptional system.

Here is a shot of the 850 Mighty Demon carb inside the enclosure used for the dyno test.

The HP Performance kit included a 66mm Holset turbo. We have run this turbo previously on the 331 and selected it in expectation of immediate boost response on the larger 408. We may select a slightly larger unit before making the 200-mph passes.

to ever reach top speed. Even though it has the power to run 200 mph (for instance), it may not have enough power to run the required 180 or 190 mph it takes to eventually see 200 mph. The cure to this potential shortcoming is the additional torque offered by turbo.

Since we had such good luck with the turbo kit from HP Performance on our 1,000hp 331, not to mention all of the turbo testing performed on the motor at the lower boost levels (even in carbureted form), we decided to enlist their help on the 200-mph project as well.

With our turbo kit selected, we needed an appropriate turbo motor to install it on. Knowing torque production was the name of the game (to provide sufficient power to overcome the

fourth/fifth gear split in our Tremec TKO II), we chose to upgrade the displacement of our project motor. If there is one thing I have learned from all the years of running top-speed events it is always choose a bigger motor over a small one.

While this sounds rudimentary, there is some hidden logic that goes beyond the absolute ability to produce more power. Unlike the exercise with our 1,000hp motor 331, we were not looking to produce maximum power with this buildup. In fact, I have on more than one occasion had to remind the guys at HP Performance we did not need to make 800, 900 or 1,000 hp despite the choice of a motor capable of doing so. They kept trying to talk me into installing a 1,000hp turbo on what need only be a 650hp motor. Sorry guys, but the buildup is actually application driven and that application is not maximum power production. We need only produce enough power to achieve the goal of 200 miles per hour, but we need to have the motor alive and well after doing so.

My choice of a 408-cubic-inch D.S.S. stroker was made not for maximum power potential, but rather to allow the required power to be produced as easily (and reliably) as possible. There is no doubt it is possible to make enough power to coax a Mustang to 200 mph using a smaller 302, but it would take a fare bit of boost pressure. In my experience, high boost pressures tend to complicate matters. A big motor with low boost is always preferred to a small motor with big boost when it comes to reliability.

My most impressive run at the Silver State Open Road Race (and to my knowledge still the only time ever won overall by a Mustang) was with the boost pressure set at just 7 psi. Running 7 psi I knew the motor would go out and run at wide open throttle all day long. Think about that statement. How many people do you know who have put their foot to the floor of their supercharged Mustang and left it pinned there until it ran out of gas? Big power and even big top speeds are great, but all must be accomplished with absolute reliability. The extra 5–6 psi required by the smaller 302 to equal the power of the larger 408 comes with a price. That price is reliability, not to mention extra heat and the attending potential for detonation.

Given our goal was a particular power output (let's call it 550 wheel hp), the best (most reliable) way to achieve that goal was with a large displacement motor. By now you should recognize the fact I am a firm believer in the simple concept a good turbo motor starts out as an efficient normally aspirated motor. In fact, the better (more powerful) the normally aspirated combination, the easier it is to achieve your desired power goal once you add the turbo to the equation. Knowing this we did everything we could to produce an efficient stroker motor before installing the turbo kit from HP Performance. The 408 began life as a stroker short-block supplied by the stroker experts at D.S.S.. Ford Racing supplied a Sportsman block to serve as the basis for the D.S.S. buildup.

To the stout block D.S.S. added the necessary 4.00-inch stroker crank, forged connecting rods and forged (low-compression) dished pistons. To further enhance the stout (two-bolt) short-block, D.S.S. also included one of its billet main stud girdles and matching windage tray. To this we added a Canton oil pan and matching pickup, a high-volume oil pump and ARP pan bolts. In fact, ARP supplied the fasteners for the entire buildup including the 1/2-inch head studs.

With our stroker short-block ready, we turned our attention to maximizing airflow. The first step was choosing a suitable camshaft. Eventually we decided on one of Comp Cams Street Roller profiles. The XR286R Street Roller cam was chosen for two reasons, the first of which was the fact we wanted to be sure we could rev the motor to 6,500 rpm should it become necessary to lessen the drop between fourth and fifth gears. In addition to stretching fourth gear, we also wanted the cam to help the motor produce exceptional power while working well with the intended turbo kit.

Having run the XR286R previously on our 331 turbo, we knew it was very turbo friendly. We also knew the cam worked equally well in normally aspirated guise. We hoped the cam combined with the single-plane Victor Jr./Spyder intake (converted by CHP for EFI use) would allow the motor to

Innovative Turbo supplied a suitable wastegate for our 408. Note the mounting position inline with the exhaust flow. Once again, attention to detail helps improve the effectiveness of the overall system.

MSD supplied billet distributors for our carbureted (dyno) and EFI (in car) applications.

produce peak power near 6,000 rpm, but rev cleanly past if necessary. It is very important to match the operating range of the cam with the intake manifold. Our XR286R roller cam and Victor Jr. intake were ideally suited to optimize power in the same rpm range.

Next on the list was a set of cylinder heads, the aforementioned AFR 205 heads. After running them in the 393 stroker, we installed them on the 408 and ran the motor on the dyno the very next day. With seven sets of cylinder heads and then a complete 408 turbo buildup, it was quite a busy week, but the results were well worth the effort. The AFR 205 heads featured 205cc intake ports that flowed over 300 cfm.

Even more impressive were the exhaust ports, which flowed 244 cfm, giving the AFR heads an intake-to-exhaust flow relationship of 81 percent. Credit for the impressive flow goes to the port shape and trick CNC porting. Even the combustion chambers were given the CNC touch. The heads featured a 2.05/1.60 valve combo and were set up (by AFR) with a spring package suitable for our Street Roller cam. The AFR heads were installed using ARP head studs and a set of Fel Pro 1011-2 head gaskets.

The motor was first run in carbureted trim, as our all-mighty deadline did not allow us to run the

Running just 7 psi of boost, the 408 produced 725 hp and 714 lb-ft of torque. Way down at 2,600 rpm, the turbo 408 thumped out 650 lb-ft of torque.

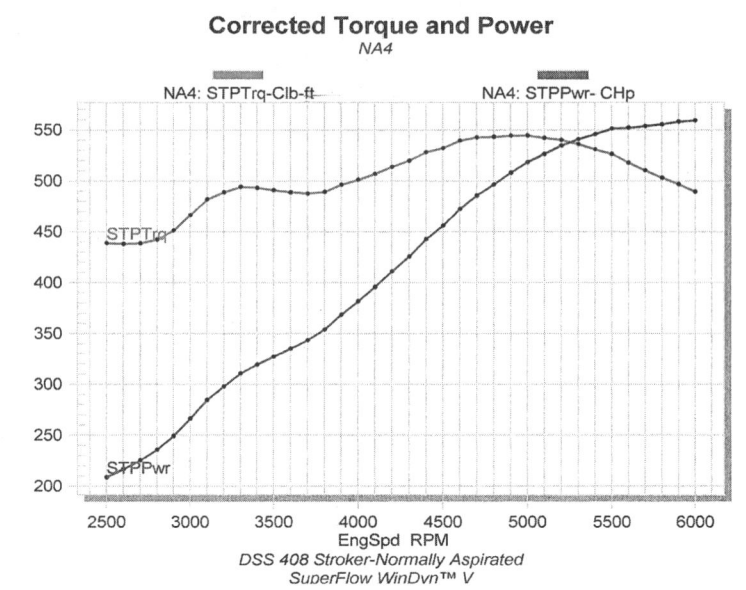

Corrected Torque and Power
NA4

NA4: STPTrq-Clb-ft   NA4: STPPwr- CHp

EngSpd RPM

DSS 408 Stroker-Normally Aspirated
SuperFlow WinDyn™ V

NORMALLY ASPIRATED D.S.S. 408—HP & TQ

Adding the right combination of cam, heads and intake to the D.S.S. 408 short-block resulted in one impressive low-compression stroker motor. The AFR 205 heads and Edelbrock Victor Jr. intake provided plenty of breathing while the XR286R roller cam allowed the motor sufficient cam timing to produce peak power at 6,000 rpm. Not shown on this graph is the ability to rev cleanly to 6,600 rpm where the 408 still produced over 550 hp. Though equipped with a healthy roller cam and single-plane intake manifold, the 408 still managed to produce 439 lb-ft of torque all the way down at 2,500 rpm. That is more torque than a normally aspirated 302 could hope to make at its peak. It is this exceptional low-speed torque that helps make for snappy turbo response.

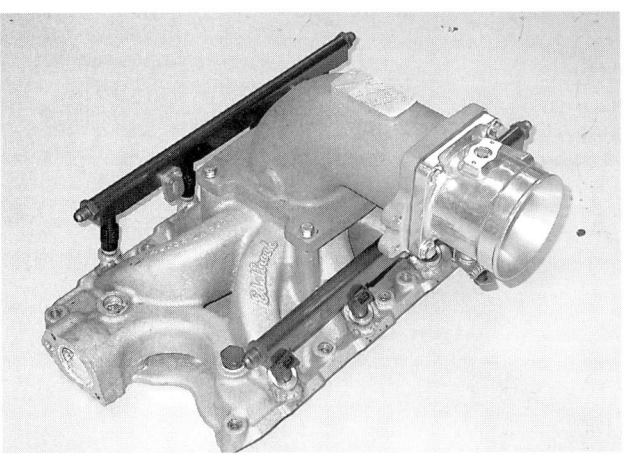

Though run in carbureted form on the engine dyno, the motor will be equipped with the factory Ford EEC-IV fuel injection in the car. The intake choice will be the same, as we selected a CHP Spyder (converted Victor Jr.) for our 408.

motor on the engine dyno using the programmable F.A.S.T. fuel-injection system. We installed a carbureted Victor Jr. intake and matching 950 HP Holley carb along with a set of Hooker 1 3/4-inch Super Comp headers. Additional components included a set of 1.6-ratio Crane Gold roller rockers, a CSI electric water pump and MSD billet distributor.

After a short break-in procedure, the carbureted 408 produced exceptional normally aspirated power. After some minor jetting and setting the total timing at 32 degrees, the D.S.S. 408 produced 562 hp at 6,000 rpm and 545 lb-ft of torque at 5,000 rpm. Though equipped with a healthy roller cam, the 408 produced some pretty impressive low-speed torque numbers. We pulled the motor all the way down to 2,500 rpm where the 408 thumped out 439 lb-ft. The combination of the cubic inches and a healthy cam/intake combo allowed us to produce an exceptionally broad power band.

With our fuel injection MIA we had no choice but to install the HP turbo kit on the carbureted 408. Having run a carbureted/turbo motor previously, we knew just what went into making them work on the dyno. Basically all that was necessary was to enclose the carburetor in a Paxton/Vortech carb bonnet and boost reference the fuel pressure regulator. This way the carb bowls would always have fuel pressure feeding them that exceeded the pressure in the carb enclosure. We swapped out the Holley 950 HP for a Barry Grant 850 Mighty Demon. The carb was chosen for its previous use on a supercharged application.

We were hoping to minimize the tuning as jet changes required (a time-consuming) removal of the carb from the bonnet. The HP Performance turbo kit was installed on the 408 sans intercooler, as the dyno didn't lend itself to the front-mounted core. Since we planned on running just 7 psi, we knew this wasn't going to present a problem, but hedged out bet with a load of 100-octane race fuel. Our 351 W Turbo kit from HP Performance included a turbo upgrade from the 60mm turbo run on the

302s to the 66mm unit. The larger turbo was deemed necessary to supply sufficient airflow to the larger 408.

For our custom-carbureted application, the turbo was rotated 180 degrees on the mounting flange. The turbine housing was oriented toward the passenger-side fenderwell and the compressor housing toward the driver's. This allowed the compressor discharge to be connected directly (with a straight section of tubing) to the 90-degree elbow used on the Paxton car bonnet.

Once everything was hooked up and cooperating, we began the turbo testing. The turbocharged 408 proved to be very powerful, peaking at 725 hp at 6,000 rpm and 714 lb-ft at 4,900 rpm. Note the power peaks occurred at the same engine speeds as the normally aspirated combination. Though a tad on the small side for this large-displacement stroker, the 66mm turbo provided impressive boost response. Loading the motor down to 2,600 rpm, the turbo 408 produce near maximum boost (6.5 psi) along with a tire shredding 681 lb-ft of torque.

Things will likely improve once we install the turbo 408 in the car and are able to have complete control over the fuel-injection system not to mention adding the cooling capacity of the air-to-air intercooler. We plan on a full tuning regimen at PowerTrain Dynamics once the motor is installed. We'll make that 200-mph number yet!

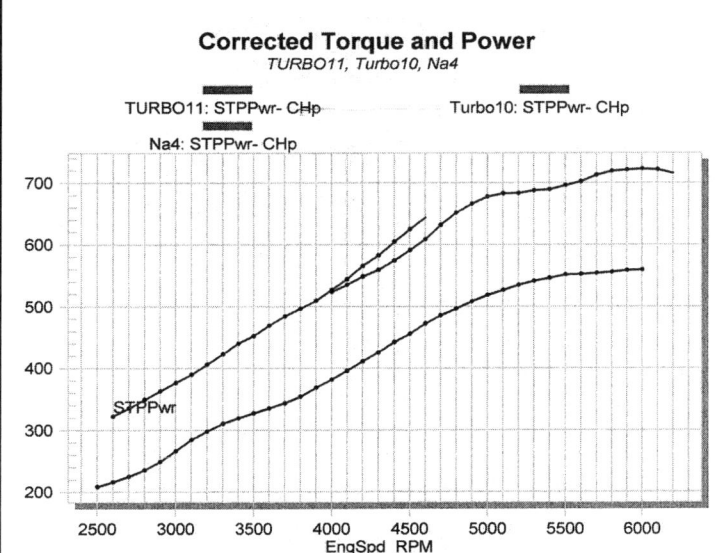

Corrected Torque and Power
TURBO11, Turbo10, Na4

TURBO11: STPPwr- CHp   Turbo10: STPPwr- CHp
Na4: STPPwr- CHp

DSS 408 Stroker-Normally Aspirated vs HP Turbo (7 psi)
SuperFlow WinDyn™ V

NA vs. Turbo 408— Horsepower

Adding a quality turbo kit to a healthy normally aspirated motor is an easy route to serious power production. Despite installing a relatively small turbo (done for impressive boost response on the street), the 408 produced 725 hp at just 7 psi. We know that more peak (and mid range) power was available had we elected to install a larger compressor housing (something near 72 mm), but the combination made more than enough power to propel our Mustang to a 200-mph top speed. In fact, if all goes well, we will probably only need to run 5 psi to accomplish our feat. Imagine that, a 200-mph Mustang running just 5 psi of boost.

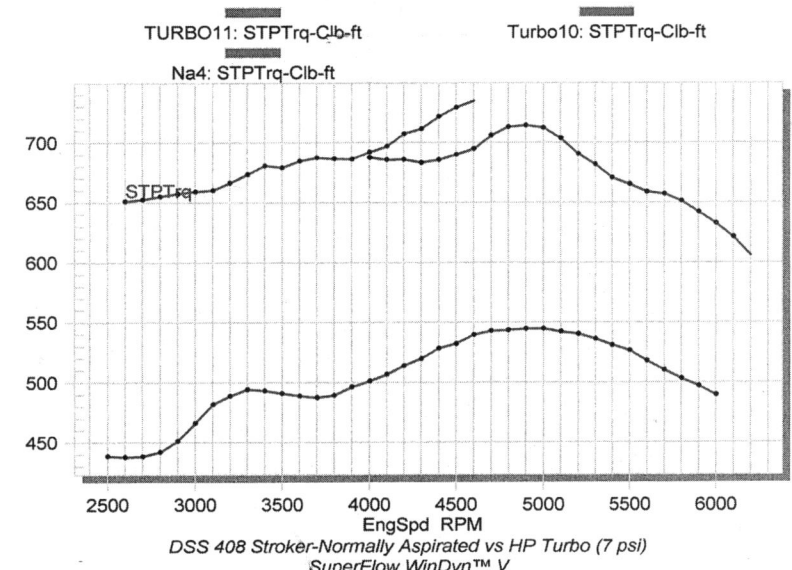

Corrected Torque and Power
TURBO11, Turbo10, Na4

TURBO11: STPTrq-Clb-ft   Turbo10: STPTrq-Clb-ft
Na4: STPTrq-Clb-ft

DSS 408 Stroker-Normally Aspirated vs HP Turbo (7 psi)
SuperFlow WinDyn™ V

NA vs. Turbo 408— Torque

We were very impressed by the torque production of the normally aspirated 408, but check out the torque production of the turbo 408. Adding the 66mm turbo from HP Performance allowed the 408 to thump out a very Power Stroke diesel-like 650 lb-ft at just 2,500 rpm. We were right about the torque of the 408 offering instant turbo response, as the 66mm turbo showed 6.5 psi of boost at just 2,600 rpm. If anything, we suspect the combination may be too responsive, offering more torque than our street tires can hope to handle. Before installing the motor in the car, we may take HP up on their offer to upgrade the turbo. The trade should help improve mid-range and high-rpm power with a slight trade off in boost response.

SOURCES

Accufab
1514 B. East Francis
Ontario, CA 91761
909/930-1754

AFR
10490 ILEX Ave.
Pacoima, CA 91331
818/890-0616
www.airflowresearch.com

ARP
531 Spectrum Circle
Oxnard, CA 93030
805/278-7223

Canton Racing Products
232 Brandford Road
North Branford CT, 06471
203/481-9460
www.cantonracingproducts.com

Coast High Performance
2555 W. 237th St.
Torrance, CA 90505
310/784-1010
www.coasthigh.com

Comp Cams
3406 Democrat Rd.
Memphis, TN 38118
901/795-2400
www.compcams.com

Crane Cams
530 Fentress Boulevard
Daytona Beach, FL 32114
382/252-1151
www.cranecams.com

D.S.S. Racing
3550 Stern Ave.
St. Charles, IL 60174
630/587-1169
www.dssracing.com

Fel Pro
One Equation Boulevard
Ashland, MS 38603
662/224-8972
www.federal-mogul.com

Ford Racing Performance Parts
15021 S. Commerce Dr., Ste. 200
Dearborn, MI 48120
800/367-3788
www.fordracingparts.com

HP Performance Inc.
301 4th St.
Roswell, NM 88201
505/623-2555

Innovative Turbo
845 Easy Street Unit 102
Simi Valley, CA 93065
805/526-5400
www.innovativeturbo.com

MSD Ignition
1490 Henry Brennan Dr.
El Paso, Texas 79936
915/855-7123
www.msdignition.com

Vortech Engineering
1650 Pacific Ave.
Channel Islands, CA 93033
805/247-0226
www.vortechsuperchargers.com

A Hardcore 427ci engine is ready to be shipped to a customer. This bad boy produces 525 hp and 505 lb-ft of torque. The engine is complete from carb to pan— just add headers, starter, and accessories. The 9.5:1 compression ratio allows the use of pump gas.

Ford rolled the last pushrod Mustang off the assembly line over a decade ago. What that means is most of you are running around with a 5-liter engine that is well broken in. And who are you kidding—your moneymaker under the hood has had a hard life on the streets of Anytown, USA. For many Stangs, it's about that time to dump the factory 302 and get a fresh bullet. But sometimes things are easier said than done—or are they?

One may ask a lot of questions as the search begins for a suitable replacement for his or her beloved 302, such as, "When will the engine be ready?" "Will it make power?" and "What size is best for me?"

Often the "When will it be ready?" question is the most troubling of all. Some engine builders move at a rate that makes construction of the great pyramids seem speedy by comparison.

One option for engine shoppers is to turn to the increasingly more popular crate-engine segment of the Ford aftermarket. Ordering such a powerplant is a great alternative to a custom engine. It may not be made to your exact specifications, but the reliability, affordability, and consistency makes this genre appealing.

One such new company is Bill Mitchell's World Products. Long known for its engine components, including cylinder heads and engine blocks, as well as a vast line of Chevy crate engines and parts, about a year ago the company designed and started producing small-block Ford crate engines. World Products delivers motors that not only make great horsepower, but are also backed by a two-year, 24,000-mile warranty and are available from any World distributor. We'll remind you that most of these mail-order monsters make well in excess of 400 hp—even over 500 hp. Please take a moment to pinch yourself so you know that you are not dreaming over the fact a turnkey product with a warranty makes that kind of horsepower.

World Products offers several options, from basic, stroked

Man-O-War blocks are used in all crate-engine applications. There are several versions of this block for the different small-block Ford deck heights. A clean sheet of paper was used when designing the Man-O-War blocks. Bill Mitchell and his design team spent a considerable amount of design time with a prominent Nextel Cup race team. They reverse-engineered the current line of Ford blocks and solved many of its problems, while improving on existing features.

The "C" blocks (optional for 8.2- and 8.7-inch-deck heights) feature a revised distributor and oil pump location. Their new location is the same as a 351 block and enables a larger stroke crankshaft to be used. This allows the Man-O-War block to house larger-than-normal engine sizes. Building a 375-cube engine from an 8.2-inch-deck block is now possible.

Here is a close-up of the crankshaft for the 351ci engine. As you can see, it needed to be shaved a bit in order to clear different spots on the block and oil pump. World Products uses Eagle cranks in all of its stroker engines.

Splayed four-bolt caps made from billet steel are standard features on the Man-O-War.

To ensure prolonged life and reliable service, all crankshafts are balanced in-house. Here, the Eagle crank that was clearanced in the previous photo goes for a whirl on the machine. World Products has an enormous room filled with all sorts of machines to handle all aspects of engine building. Everything is done under one roof. In fact, the company has two of every machine, except for engine dynos—it has three of those.

Engine builder Mike McIntyre lowers the crank into place. This particular arm has a 3.500-inch swing, and when combined with the 4.000-inch bore, it yields a 351ci displacement.

A one-piece rear main seal is utilized. This helps prevent leaks, and was introduced on small-block Ford engines around 1982 or 1983.

302-style engines to track-only 427 combinations. They are all based around the company's Man-O-War engine block. Thanks to a new design and machining process, the Man-O-War block is both affordable and strong—two words that aren't often joined together in this industry. We ventured to World Products, located in Ronkonkoma, New York, to watch the buildup of two of its crate engines. Mike McIntyre, veteran engine builder, assembled two World Class engines—the 351ci

(302-based engine) and the company's 427ci version.

The Man-O-War block is the culmination of a 2 1/2-year investment in time and money. Mitchell and his staff worked alongside a prominent NASCAR Nextel Cup team when designing this piece. They started with a clean sheet of paper and worked from there. By using C.A.D. engineering software, the World Products crew came up with what they feel is a far superior engine block than anything else on the small-block Ford market. They improved the oiling system and head-seal problems, and added extra webbing where it was needed and thicker cylinder walls, among other things.

"Using the C.A.D. design we were able to 'remove' the oiling system from the block," Mitchell says. "We could then see where the problems were. The Man-O-War blocks route the oil the way it is suppose to be routed." This longtime enthusiast and engine parts designer went on to say that the small-

If you look closely, you can see the gear markings on the sprocket. This is to help get the cam degreed properly. It will usually help the engine builder get to within 1 degree of where it is supposed to be.

The type of camshaft varies between each engine combination. As for the specs on this one, the World Products crew wasn't saying. What we do know is it was designed for street use and provided a nice, choppy, yet subtle, idle.

Products uses the 8.2- and 9.5-inch blocks in its crate-motor lineup. This is to ensure they are a direct drop-in when used in Mustangs and other Ford vehicles.

World Products' catalog shows 11 different small-block Ford crate engines. Horsepower and pricing are what separates each part number. The World Class lineup is on the lower end of the scale, while the Limited Edition and Drag Racing engines are on the higher end. The smallest cubic-inch offering checks in at 351 ci, while the largest engine is a whopping 460 ci (it is dubbed a Limited Edition model). The middle-of-the-road engine has been rightfully nicknamed "Hardcore."

Once an engine is assembled, it's delivered to one of the three DTS engine dynos at the World Products facility. Each and every engine is run to evaluate power production, ensure there are no leaks, and that everything is operating properly. This procedure is done so all crate engines are delivered trouble-free and making the advertised horsepower.

How can you beat that? World Products' crate engines eliminate the guesswork and are designed to log some serious street miles. They even have a guarantee to back up those claims. And ordering one is as easy as opening up a catalog.

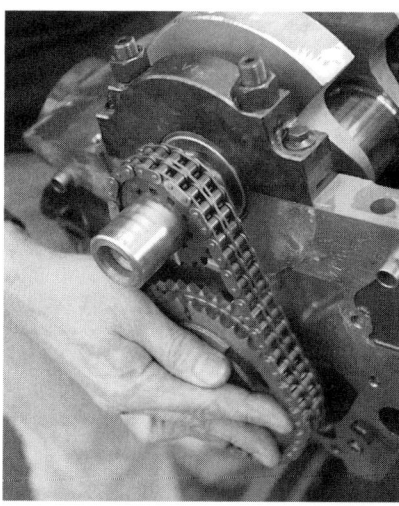

McIntyre installs a Hardcore double-roller timing chain. This type of chain is used in all World Products crate engines.

Eagle rods are used in all applications. Like most of the rotating assembly, sizes vary depending on the application.

block Ford engines have a lot of features, such as the camshaft location, the distributor location, and especially the various deck heights that were a nice departure from the Chevy/GM way of doing things.

With a sturdy foundation to work on, the World Products staff could use larger bores and help increase the displacement of their engines without sacrificing driveability and longevity. There is actually a 375ci engine available that is still based on an 8.2-deck engine block, dubbed the "C block." That size cubic inch was unheard of from a short-deck Ford—until now. Thanks to the bores being able to handle a 4.125-inch size and the block handling a 3.500-inch crank, 375 ci aren't a problem. The largest sized engine checks in at 427 ci and that too runs a 4.125 bore without issues. This all stems back to the Man-O-War being the ultimate starting point.

The engine blocks may be a little heavier than a stock casting, but durability and better oiling and cooling designs far outweigh the physical comparison—no pun intended. The weight comes from thicker iron and the fact that splayed four-bolt billet steel main caps are standard. A stud girdle is definitely not required in this application. All Ford small-block deck heights are available—8.2 inches, 8.7 inches, 9.2 inches, and 9.5 inches. World

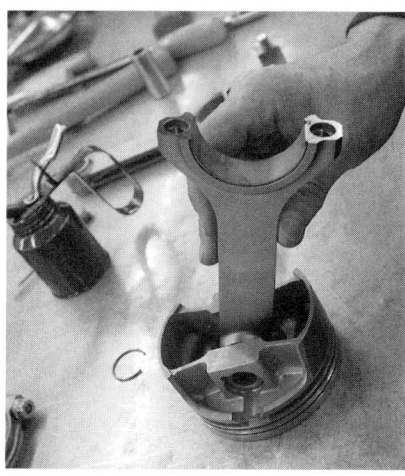

Mahle pistons are connected to the Eagle rods, and will hopefully stay that way for a long time. The pistons have a coating on the side skirt.

The piston-rod combo is slid down into the bore.

This Eagle rod is bolted together using ARP L19 rod bolts.

All camshafts are degreed in every engine buildup. The engine assemblers go through a rigorous checklist to ensure all components are installed and fit properly. Degreeing the camshaft is just one of those mandatory processes.

A Melling oil pump is used to keep the engine properly lubricated. A unique feature to the Man-O-War blocks is a new oiling system. Mitchell informed us that because this is an aftermarket block that is not produced in the hundreds of thousands each year, they could do things a bit differently than the OEMs. The Big Three have more than just performance on their minds when designing parts. The oiling system is just one of the many unique features built into these blocks.

World Products provides a Canton or Milodon oil pan with each of its engines. The pan features a double sump to clear most K-members, but be sure to double-check it before dropping it into your Mustang.

A Pro Race internal balancer is used in all applications.

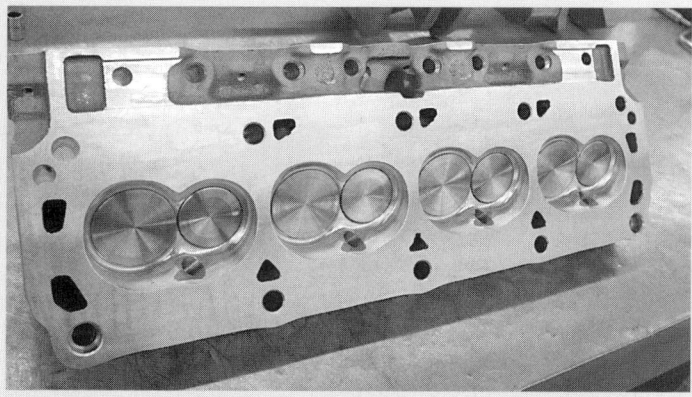

There are two choices for cylinder heads-—aluminum or steel. The castings are World Products Windsor Sr. cylinder heads.

Manley valves are used in the Windsor Sr. cylinder heads. The intake valves measure 2.020 inches and the exhaust valves check in at 1.60 inches.

McIntyre lowers the head into place. A Fel-Pro head gasket is used, as well as ARP studs, to seal and fasten the head to the block. A point of note here is the Windsor heads do not use the extra head-bolt provisions on the Man-O-War block.

Scorpion roller rocker arms, Hardcore rocker arm studs, Hardcore pushrod plates, and one-piece Hardcore pushrods make up the valvetrain in all the crate engines.

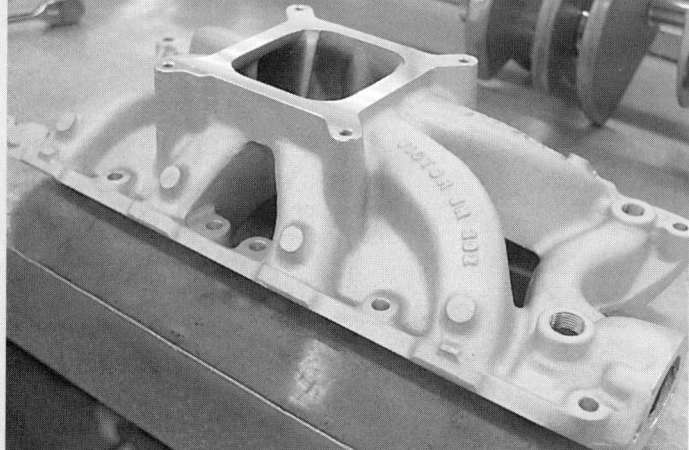

Edelbrock single-plane intake manifolds help these engines breathe deeply.

A Mallory HEI distributor sends spark to all eight holes. Even the spark plug wires and spark plugs are included in the package.

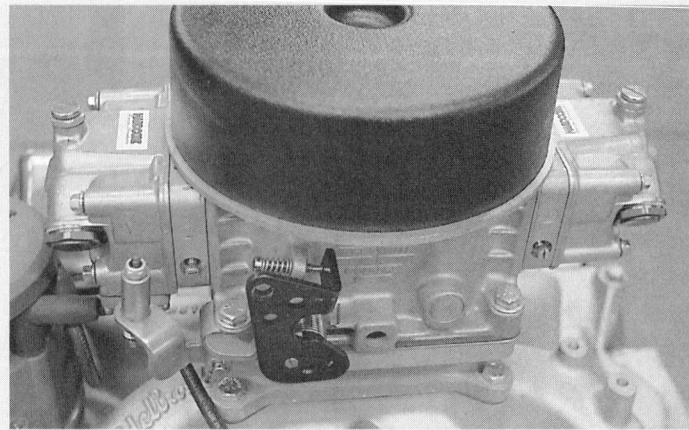

World Products uses specially designed carburetors for its engines. Shown at left is a Hardcore 870-cfm (4150-style) carb, while at right is a Hardcore 1,050-cfm (4500-style) carburetor.

Three DTS dynos are set up at World Products. They are used to test each engine before it is shipped. The staff evaluates horsepower and torque output to ensure each engine is within spec. World Products is also home to the Popular Hot Rodding Engine Masters Challenge.

WORLD PRODUCTS SMALL-BLOCK FORD CRATE ENGINES

World Class Engines

CID	Horsepower	Torque	Block	Heads
351	395	385 lb-ft	8.2-in Man-O-War	Cast Iron
351	415	405 lb-ft	8.2-in Man-O-War	Aluminum
375	425	425 lb-ft	8.2-in Man-O-War	Cast Iron
375	455	445 lb-ft	8.2-in Man-O-War	Aluminum
427	475	465 lb-ft	9.5-in Man-O-War	Cast Iron
427	495	485 lb-ft	9.5-in Man-O-War	Aluminum

Hardcore Engines

CID	Horsepower	Torque	Block	Heads
427	500	490 lb-ft	9.5-in Man-O-War	Cast Iron
427	525	505 lb-ft	9.5-in Man-O-War	Aluminum

Limited-Edition Engine

CID	Horsepower	Torque	Block	Heads
460	575	550 lb-ft	9.5-in Man-O-War	Aluminum

Note: World Class, Hardcore, and Limited Edition engines are designed to work on pump gas and are backed by an independent insurance company for two years or 24,000 miles.

Drag Racing Engines

CID	Horsepower	Torque	Block	Heads
427	600	565 lb-ft	9.5-in Man-O-War	Cast Iron
427	625	575 lb-ft	9.5-in Man-O-War	Aluminum

Note: All Drag Racing Engines are designed to run on racing fuel and are not for street use.

SOURCE
World Products
51 Trade Zone Ct.
Ronkonkoma, NY 11779
631/737-0372
www.theengineshop.com

MAN-O-WAR ENGINE BLOCKS AT A GLANCE

Material: Cast Iron
Deck Heights (in inches): 8.200, 8.700, 9.200, and 9.500
Deck Thickness: 0.600-in minimum
Cam Bearings: Stock, can be bored to 60 mm
Lifter Bores: Cross-feeds between pairs to maintain pressure
Bore: 3.990-in (bore to 4.000 in) or 4.115 in (bore to 4.125 in)
Cylinder Head Attachment: Two extra bolts per cylinder*
Maximum Bore: 4.200 in (470 ci with max bore/max stroke combination)
Waterjackets: Expanded for better cooling
Lubrication: Priority main oiling system
Filtration: Integral mount for spin-on filter
Oil Pan Rails: Stock Ford
Crank Clearance: Can use up to 4.25-in crank in tall-deck blocks
Main Caps: Splayed billet steel
Oil System Note: Features 0.500-in feed and has bosses for dry-sump setup
   *Note: The two extra head bolts are not required to be used. They have been incorporated into the casting for future use with newly designed cylinder heads. World Products and Edelbrock are just two companies that we know of that have heads forthcoming with the extra bolts. This will help seal better in harsh conditions, like what is present when using nitrous or forced induction.

Chapter 20
MODULAR MOTOR CAM TIMING

Six Degrees of Separation

Text and Photos By Richard Holdener

One thing we've discovered about the OHC modular motors is that they are highly sensitive to cam timing. Unfortunately, the production tolerances are such that the cam timing can be off significantly from the factory-prescribed settings. This is especially the case with the Four-Valve motors, as they seem even more likely than their Two-Valve counterparts to be burdened with inaccurate timing. The most probable scenario is that the right and left bank cam timing differs, causing one side to be either advanced or retarded relative to the other. Both may be off from the factory specs, but the real key is that one side will produce significantly less power than the other. The key to balancing power production is dialing in the cam timing.

This procedure takes time and know-how, but the results can be significant. Dialing in the cam timing on one of our 4.6 test motors (a mild one at that) was worth as much as 22 hp. Balancing the cam timing side to side on a 4.6 Cobra motor was worth a solid 12–15 hp, from 3,500 rpm to 6,500 rpm.

Dialing in the cam timing on the 4.6 is obviously much easier with the motor out of the car, but it can be accomplished with the engine in place. This info will likely be more useful for engine builders planning on installing new or rebuilt motors, but rest assured that cam timing is more critical on the mod motors than the previous-generation 5.0 engines.

A word of advice: Determine whether your mod motor suffers from this cam-timing malady. Performing a compression test (something easily accomplished by even a backyard mechanic) will show if one bank of cylinders produces a higher compression reading than another. The adjustment procedure will be much easier with a set of adjustable sprockets, such as the ones shown from Fidenza, but we performed our testing using modified stock sprockets. The best method (short of using a degree wheel and dial indicator) would be to adjust the cams until both banks produce the same cranking compression. The Two-Valve 4.6 motor ran best after we balanced the compression to match the lowest reading (both at 155 psi). In the case of our 4.6 Four-Valve motor, the best power came after the compression of both banks produced the highest reading (200 psi). Once we have adjustable cam sprockets, we should be able to dial in the cam timing with the cranking compression gauge.

Though the factory cam sprockets feature a keyway to positively locate the cam, it is possible to file down the keyway or remove it all together to adjust the cam timing.

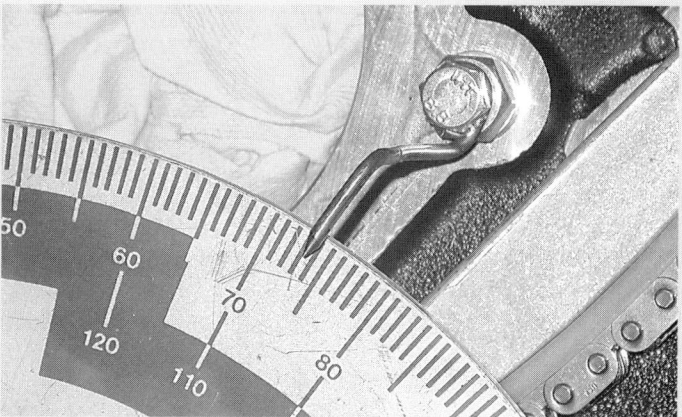

It is necessary to use a degree wheel and dial indicator to properly check the cam timing.

On this SHM 4.6 Two-Valve motor, dialing in the cam timing resulted in some serious power gains.

We found that a compression test was also a good indicator as to the balance (side to side) of the cam timing.

Balancing the cam timing side to side on the SHM 4.6 Four-Valve motor resulted in a sizable chunk of power.

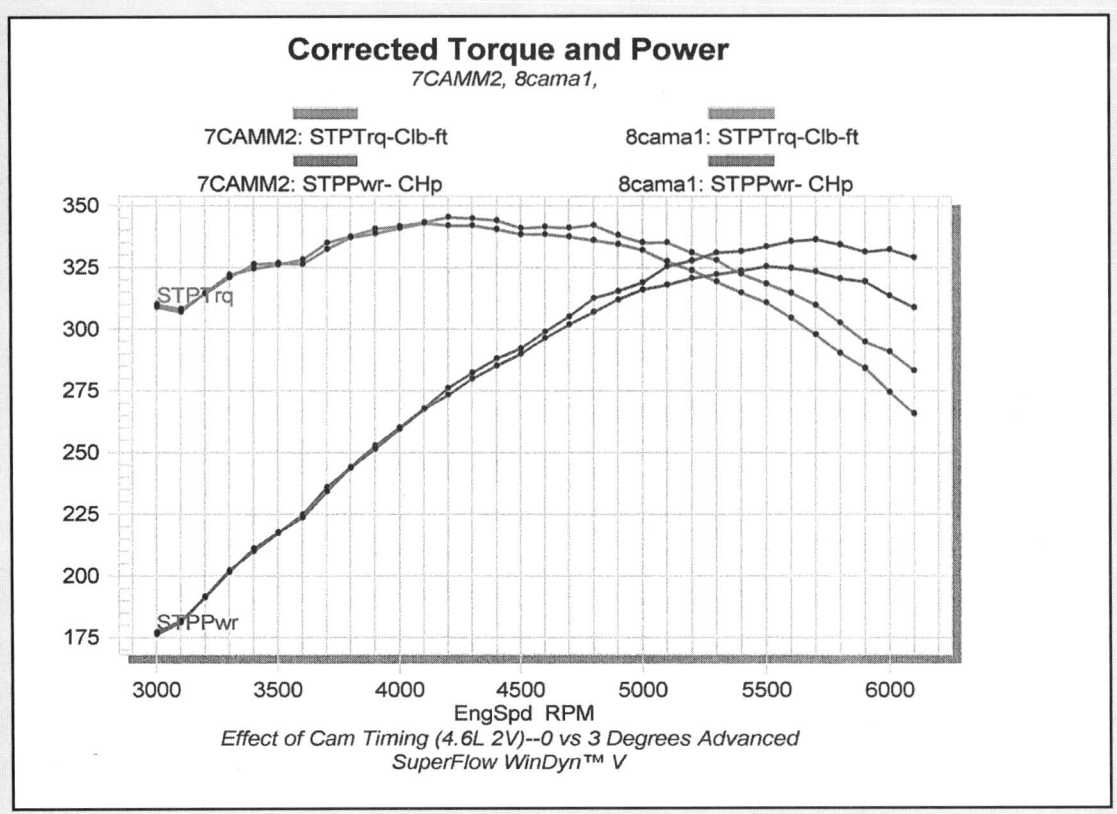

Corrected Torque and Power
7CAMM2, 8cama1,

7CAMM2: STPTrq-Clb-ft   8cama1: STPTrq-Clb-ft
7CAMM2: STPPwr- CHp   8cama1: STPPwr- CHp

Effect of Cam Timing (4.6L 2V)--0 vs 3 Degrees Advanced
SuperFlow WinDyn™ V

EFFECT OF CAM TIMING (4.6L 2V)
0 VS. 3 DEGREES ADVANCED

This test was performed on a 4.6 2C PI motor supplied by Sean Hyland. The PI motor was equipped with stock compression, heads, and intake. The only modifications to the motor were a set of Kooks 1⅝-inch headers and a set of Comp XE262H (non-PI) camshafts. While performing the cam swap, we noticed one of the cam sprockets had been altered to allow it to be adjusted slightly. By filing down the keyway, the sprocket could be adjusted over a small range.

Unfortunately, we did not pay attention to the position when we removed it during the cam swap. This mistake allowed us to demonstrate the effect of changes in cam timing, as it was necessary to dial in the cam position to optimize the power output. When first installed, the right (passenger side) cam was equipped with a standard (nonadjustable) cam sprocket. The left (driver side) was equipped with the adjustable sprocket. We attempted to run the motor with the left cam advanced 3–4 degrees relative to the right cam. A compression check on each bank of cylinders resulted in readings of 175 psi for the left (advanced) cam and just 155 psi for the right (0 position) cam.

Though you might be inclined to advance the right cam to equalize the (higher) cranking compression offered by the left cam, the reverse actually turned out to be the correct action. Retarding the left cam back to the 0 position (so both sides produced 155 psi cranking compression) resulted in a significant power gain (as much as 23 hp). Obviously, dialing in the cam timing on the 4.6 Two-Valve is critical to maximum performance.

METRIC CUSTOMARY UNIT EQUIVALENTS

Multiply:	by:	to get:	Multiply by:	to get:
LINEAR				
inches	x 25.4	= millimeters (mm)	x 0.03937	= inches
feet	x 0.3048	= meters (m)	x 3.281	= feet
yards	x 0.9144	= meters (m)	x 1.0936	= yards
AREA				
inches^2	x 645.16	= millimeters^2(mm^2)	x 0.00155	= inches^2
feet^2	x 0.0929	= meters^2(m^2)	x 10.764	= feet^2
VOLUME				
quarts	x 0.94635	= liters (I)	x 1.0567	= quarts
gallons	x 3.7854	= liters (I)	x 0.2642	= gallons
feet^3	x 28.317	= liters (I)	x 0.03531	= feet^3
feet^3	x 0.02832	= meters^3(m^3)	x 35.315	= feet^3
fluid oz	x 29.57	= milliliters (ml)	x 0.03381	= fluid oz
MASS				
ounces (av)	x 28.35	= grams (g)	x 0.03527	= ounces (av)
pounds (av)	x 0.4536	= kilograms (kg)	x 2.2046	= pounds (av)
FORCE				
ounces-f(av)	x 0.278	= newtons (N)	x 3.597	= ounces-f(av)
pounds-f(av)	x 4.448	= newtons (N)	x 0.2248	= pounds-f(av)

TEMPERATURE

Degrees Celsius (C) = 0.556 (F – 32)   Degree Fahrenheit (F) = (1.8C) + 32

ENERGY OR WORK (Watt-second = joule= newton-meter)				
foot-pounds	x 1.3558	= joules (J)	x 0.7376	= foot-pounds
Btu	x 1055	= joules (J)	x 0.000948	= Btu
PRESSURE OR STRESS				
pounds/sq in.	x 6.895	= kilopascals (kPa)	x 0.145	= pounds/sq in
TORQUE				
pound-inches	x 0.11298	= newton-meters (N-m)	x 8.851	= pound-inches
pound-feet	x 1.3558	= newton-meters (N-m)	x 0.7376	= pound-feet
pound-inches	x 0.0115	= kilogram-meters (Kg-M)	x 87	= pound-feet
pound-feet	x 0.138	= kilogram-meters (Kg-M)	x 7.25	= pound-feet
POWER				
horsepower	x 0.74570	= kilowatts (kW)	x 1.34102	= horsepower

COMMON METRIC PREFIXES

mega (M)   = 1,000,000 or 10^6 centi(c) = 0.01 or 10^{-2}
kilo (k)   = 1,000 or 10^3 milli (m) = 0.001 or 10^{-3}
hecto (h)   = 100 or 10^2 micro (u) = (u)0.000,001 or 10^{-6}

GENERAL MOTORS
Big-Block Chevy Engine Buildups: 1-55788-484-6/HP1484
Big-Block Chevy Performance: 1-55788-216-9/HP1216
Camaro Performance: 1-55788-057-3/HP1057
Camaro Restoration Handbook ('67–'81): 0-89586-375-8/HP1375
Chevelle/El Camino Handbook: 1-55788-428-5/HP1428
How to Customize Your Chevy Silverado/GMC Sierra
   Truck, 1999-2006: 978-1-55788-526-5/HP1526
How to Hot Rod Big-Block Chevys: 0-912656-04-2/HP104
How to Hot Rod Small-Block Chevys: 0-912656-06-9/HP106
How to Rebuild Small-Block Chevy LT-1/LT-4: 1-55788-393-9/HP1393
John Lingenfelter: Modify Small-Block Chevy: 1-55788-238-X/HP1238
LS1/LS6 Small-Block Chevy Performance: 1-55788-407-2/HP1407
Powerglide Transmission Handbook:1-55788-355-6/HP1355
Rebuild Big-Block Chevy Engines: 0-89586-175-5/HP1175
Rebuild Small-Block Chevy Engines: 1-55788-029-8/HP1029
Small-Block Chevy Engine Buildups: 1-55788-400-5/HP1400
Small-Block Chevy Performance: 1-55788-253-3/HP1253
Turbo Hydramatic 350 Handbook: 0-89586-051-1/HP1051
The Classic Chevy Truck Handbook, 1955–1960,
   978-155788-534-0/HP1534

FORD
Ford Engine Buildups: 978-155788-531-9/HP1531
Ford Windsor Small-Block Performance: 1-55788-323-8/HP1323
How to Customize Your Ford F-150 Truck, 1997-2008:
   978-1-55788-529-6/HP1529
Mustang Performance (Engines): 1-55788-193-6/HP1193
Mustang Performance 2 (Chassis): 1-55788-202-9/HP1202
Mustang Restoration Handbook ('64–'70): 0-89586-402-9/HP1402
Rebuild Big-Block Ford Engines: 0-89586-070-8/HP1070
Rebuild Ford V-8 Engines: 0-89586-036-8/HP1036
Rebuild Small-Block Ford Engines: 0-912656-89-1/HP189

MOPAR
Big-Block Mopar Performance: 1-55788-302-5/HP1302
How to Hot Rod Small-Block Mopar Engine Revised: 1-55788-405-6/HP1405
How to Maintain & Repair Your Jeep: 1-55788-371-8/HP1371
How to Modify Your Jeep Chassis/Suspension for
   Offroad: 1-55788-424/HP1424
How to Modify Your Mopar Magnum V8: 1-55788-473-0/HP1473
How to Rebuild and Modify Chrysler 426 Hemi
   Engines: 978-1-55788-525-8/HP1525
How to Rebuild Your Mopar Magnum V8: 1-55788-431-5/HP1431
Rebuild Big-Block Mopar Engines: 1-55788-190-1/HP1190
Rebuild Small-Block Mopar Engines: 0-89586-128-3/HP1128
The Mopar Six-Pack Engine Handbook: 978-1-55788-528-9/HP1528
Torqueflite A-727 Transmission Handbook: 1-55788-399-8/HP1399

IMPORTS
Baja Bugs & Buggies: 0-89586-186-0/HP1186
Honda/Acura Engine Performance: 1-55788-384-X/HP1384
How to Build Nissan Sport Compacts, 1991–2006:
   978-155788-541-8/HP1541
How to Hot Rod VW Engines: 0-912656-03-4/HP103
Mitsubishi & Diamond Star Performance Tuning:
   978-1-55788-_0_ _/HP496
Porsche 911 P_____: 1-55788-489-7/HP489
Rebuild Air-_____es: 0-89586-225-5/HP1225

HANDBOOKS
Automotive Detailing: 1-55788-288-6/HP1288
Auto Electrical Handbook: 0-89586-238-7/HP1238
Auto Math Handbook: 1-55788-020-4/HP1020
Automotive Paint Handbook: 1-55788-291-6/HP1291
Auto Upholstery & Interiors: 1-55788-265-7/HP1265
Classic Car Restorer's Handbook: 1-55788-194-4/HP1194
Custom Automotive Wiring and Electrical Systems:
   978-155788-545-6/HP1545
Engine Builder's Handbook: 1-55788-245-2/HP1245
Engine Cooling Systems: 978-1-55788-425-1/HP1425
Fiberglass & Other Composite Materials Rev.: 1-55788-498-6/HP1498
High Performance Fasterners & Plumbing: 978-1-55788-523-4/HP1523
Metal Fabricator's Handbook: 0-89586-870-9/HP1870
Paint & Body Handbook: 1-55788-082-4/HP1082
Performance Ignition Systems: 1-55788-306-8/HP1306
Pro Paint & Body: 1-55788-394-7/HP1394
Sheet Metal Handbook: 0-89586-757-5/HP1757
Welder's Handbook Revised: 978-1-55788-513-5/HP1513

INDUCTION
Holley 4150: 0-89586-047-3/HP1047
Holley Carbs, Manifolds & F.I.: 1-55788-052-2/HP1052
Rochester Carburetors: 0-89586-301-4/HP1301
Turbochargers: 0-89586-135-6/HP1135
Street Turbocharging: 1-55788-488-9/HP1488
Weber Carburetors: 0-89586-377-4/HP1377

RACING & CHASSIS
Chassis Engineering: 1-55788-055-7/HP1055
4Wheel & Off-Road's Chassis & Suspension: 1-55788-406-4/HP1406
Dirt Track Chassis & Suspension: 978-1-55788-511-1/HP1511
How to Make Your Car Handle: 0-912656-46-8/HP146
How to Build a Winning Drag Race Chassis & Suspension:
   978-1-55788-462-6/HP1462
Racing Engine Builder's Handbook: 1-55788-492-7/HP1492
Stock Car Racing Engine Technology: 978-1-55788-506-7/HP1506
Stock Car Setup Secrets: 1-55788-401-3/HP1401
The Race Car Chassis: 978-155788-540-1/HP1540

STREET RODS
How to Build a 1934–'35 Chevy Street Rod: 978-1-55788-514-1/HP1514
How to Build a 1935–'40 Ford Street Rod: 1-55788-493-5/HP1493
Street Rodder magazine's Chassis & Suspension Handbook:
   1-55788-346-7/HP1346
Street Rodder's Handbook, Rev.: 1-55788-409-9/HP1409
Street Rodding Tips & Techniques: 978-155788-515-9/HP1515